Many-electron Electrochemical Processes

For further volumes:
http://www.springer.com/series/7386

Monographs in Electrochemistry

Surprisingly, a large number of important topics in electrochemistry is not covered by up-to-date monographs and series on the market, some topics are even not covered at all. The series Monographs in Electrochemistry fills this gap by publishing indepth monographs written by experienced and distinguished electrochemists, covering both theory and applications. The focus is set on existing as well as emerging methods for researchers, engineers, and practitioners active in the many and often interdisciplinary fields, where electrochemistry plays a key role. These fields will range – among others – from analytical and environmental sciences to sensors, materials sciences and biochemical research.

Information about published and forthcoming volumes is available at http://www.springer.com/series/7386

Series Editor: Fritz Scholz, University of Greifswald, Germany

Aleksandr A. Andriiko • Yuriy O. Andriyko •
Gerhard E. Nauer

Many-electron Electrochemical Processes

Reactions in Molten Salts, Room-Temperature Ionic Liquids and Ionic Solutions

Springer

Aleksandr A. Andriiko
Kyiv Polytechnic Institute
Chemical Technology Department
National Technical University of Ukraine
Kyiv
Ukraine

Yuriy O. Andriyko
Centre of Electrochemical Surface
Technology (CEST)
Wiener Neustadt
Austria

Gerhard E. Nauer
Vienna University
Institute of Physical Chemistry
Vienna
Austria

ISSN 1865-1836 ISSN 1865-1844 (electronic)
ISBN 978-3-642-35769-5 ISBN 978-3-642-35770-1 (eBook)
DOI 10.1007/978-3-642-35770-1
Springer Heidelberg New York Dordrecht London

Library of Congress Control Number: 2013932472

Printed on acid-free paper

Springer is part of Springer Science+Business Media (www.springer.com)

Dedicated to the memory of
Professor Yu. K. Delimarskiy

Preface of the Editor

Both in nature, e.g. in photosynthesis and respiration, and in industrial electrochemistry, multi-electron processes by far outnumber single-electron processes. Hence, their significance is indisputable, but this is also true for their complexity with respect to theory and experiments. The authors of this monograph have a strong background in high-temperature molten salt electrochemistry, which concerns especially the transition metals and their ions. These elements have a rich redox chemistry and their electrochemical reactions can only be understood considering rather complex multi-electron processes. The senior Ukrainian author Aleksander Andriiko is a pupil of Juriy Konstantinovich Delimarskiy (1904–1990), the former head of the Kiev school of (high-temperature) molten salt electrochemistry which flourished over many decades. Unfortunately, the scientific treasures of wisdom gathered during that period are badly accessible to Western scientists. Hence it is a very lucky constellation that the senior author Aleksander Andriiko was joined by the young Ukrainian electrochemist Yuriy Andriyko, by the way his son, and the Austrian electrochemist Gerhard Nauer to compile this monograph. High-temperature molten salt electrochemistry of transition metals and other polyvalent metals offers a great amount of relevant information, especially for the room-temperature ionic liquid electrochemistry which blossoms nowadays. Preserving and disseminating the precious knowledge and experience of people who have actively taken part in the development of molten salt electrochemistry are thus an investment in the future of electrochemistry.

Greifswald, Germany
June 2012

Fritz Scholz

Preface of the Authors

"All things run, all things change" (Heraklit)

The main ideas of this book originate from *Electrochemistry of Molten Salts*. The two of us (AA and GN) belong to the generation of electrochemists who worked in this branch of science in times of its blossoming and evidenced its rapid degradation. Once powerful, this field of chemistry had fallen in a state of deep decline till the mid-1990s.

How and why could it happen?

Alexander Andriiko, a former Soviet Union scientist who worked in this field since the end of 1970s for about 25 years, knows that the Soviet Science had been shaken in the mid-1980s. "Remember my word, fellows, this is not for good"—that was the comment relating to the situation of science given at the beginning of perestroika (1986) by his (i.e., Andriiko's) Maitre, Academician Prof. Yu. Delimarskiy. This prophecy became a reality very soon. A last nail in the coffin of his science had been thrust after Cold War was over. Indeed, the streams from the mighty river of Military Industries that fed this science—it is not a secret anymore—dried out very soon after the collapse of SU.

Hardly ever he, the former Soviet scientist, regrets about this. Freedom is great thing. However, everything must be paid in this world. The price was a severe loss of science on 1/6 part of our world.

As for the *Electrochemistry of Molten Salts*, it was lost worldwide—not in the FSU countries only. The diagram below gives an idea about the scale of this catastrophe. It represents the structure of electrochemical researches in the world through the publications in a representative international journal in 1995–1996, just a few years after these events. As we can see, Molten Salts electrochemistry took the last place among other various branches of electrochemistry. Nothing has been changed since then.

Deplorable situation—especially taking into account leading positions of Molten Salt Electrochemistry some 20–25 years ago.

All things change indeed. None of us is working in the field of pure Electrochemistry of Molten Salts anymore. A. Andriiko heads the Chair of General and Inorganic Chemistry in National Technical University of Ukraine "Kyiv Polytechnic Institute"; Gerhard Nauer, a professor of Physical Chemistry, works in Vienna State University; the youngest of us, Yuriy Andriyko, has recently received his PhD degree from Vienna Technical University for the research in a newly emerging field—electrochemistry of ionic liquids.

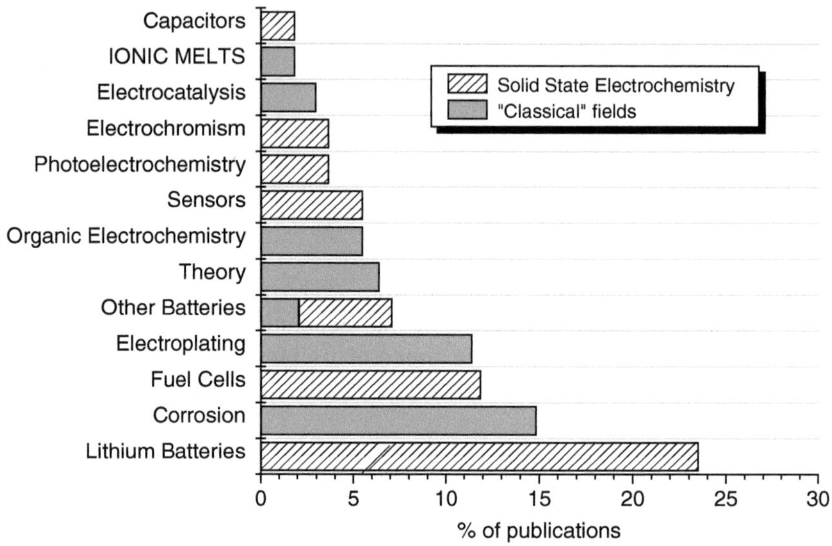

Why did we decide to abandon our routine work and spend our time for preparation of this book?

There are two reasons:

First. Now we know that the ideas are of interest not only for the electrochemistry of molten salts, but for other research fields as well. In particular, the recent results in the Electrochemistry of Ionic Liquids fit perfectly well in the developed pattern. We believe that our nature is united; we ourselves divide it conventionally for the reasons of analysis and simplification.

Second. Regretfully, our scientific knowledge is not only accumulated but can dissipate under certain circumstances. Following curious example comes to our mind. One of us (AA), being young researcher in the mid-70s and having conducted his studies on thermodynamics of electrode–electrolyte interface, came across a paper in worldwide known scientific journal announcing the "discovery of new phenomenon, reversible heat effect at first kind–second kind conductor interface, similar to the Peltier effect at a junction of two different metals". Digging deeper, he found with astonishment that this phenomenon was long ago described theoretically, verified experimentally with very high precision and published in a series of papers in *Zeitschrift fur Elektrochemie* (Lange and others, 1930s).

Who knows how many such scientific treasures were buried—possibly forever—in those bulky German "*Zeitschriften*" after Germany lost the War? It seems that we shall never get an answer to this question. However, we can predict for sure a vast cemetery of scientific ideas and results buried in papers published in Russian language in scientific journals of the USSR, who also lost this (Cold) War. Who would read them now, if they were not too wide cited even in those times of communist empire, which is now vanished?

The present book contains two kinds of the results. The first, more numerous, obtained mainly in Kiev in 1980s—the beginning of 1990s, belong to electrochemistry of "classic" high-temperature ionic melts. Being scattered in Russian language papers, they are practically unknown to Western readers. The second are recent results on the electrochemistry of polyvalent metals in ionic liquids, which, as we shall see, are in very good agreement with the first.

This book is not for a popular chemical reading. Basic knowledge in modern theoretical electrochemistry and some mathematical skills are expected from a reader, though we tried not overloading the text by formulas and equations. However, the equations are sometimes very useful for someone writing in a language that is not native for him (as it is the case for each of us). As Poincaré once aptly said, they serve as crutches for a limping man, because using a foreign language is like walking for a lame.

Thus, the book is addressed to anyone who is interested in general problems of electrochemistry—but not only for him or her. Perhaps, the book will be of use for future generation of researchers when Electrochemistry of Molten Salts revives again. We have such hope, because we believe not only in the dictum of Ancient Greeks with which we have started, but also in far more ancient wisdom of Aecclesiastus: "Nothing is new under this Moon . . ., and everything turns back in its common cyclic way".

Kyiv, Ukraine — Aleksandr A. Andriiko
Wiener Neustadt, Austria — Yuriy O. Andriyko
Vienna, Austria — Gerhard E. Nauer

Introduction

Thus, bare facts cannot satisfy us; in other words, we need science ordered, or, better to say, organized.

H. Poincaré

This book deals with the processes in electrochemical systems containing compounds of polyvalent elements. Most of the theoretical generalisations and ideas are based on the experimental data relevant to molten high-temperature systems containing complex fluorides. For a long period of time, such systems were of significant practical interest for technology of nuclear materials, electrometallurgy of refractory and non-ferrous metals and other applications [1–3]. Thus, the systems were investigated extensively throughout the world. As a result, a vast amount of experimental data has been accumulated.

Analysis of these data gives the impression of chaos and disorder. The results and conclusions of different authors are often controversial, and the data do not fit any logical pattern (possibly, that is the reason why no comprehensive reviews or monographs are available in this field). Some experimental phenomena did not receive satisfactory theoretical explanations, such as, for example: the formation of metal–salt cathode deposits, the breakdown of the industrial electrowinning of Al–Si alloys and the anode effect.

The general feature of such systems is that the electrochemical processes are many-electron ones since compounds of polyvalent elements are, typically, changing their oxidation states by more than one.

The first main idea of this work is to refuse the assumption of possible one-step transfer of several (more than one) electrons in one elementary electrochemical act and to consider any real many-electron process as a sequence of one-electron steps. Although this idea is not new (it follows immediately from quantum theories of electron transfer [4]), it is not followed consistently in research practice. The reason is that a number of significant problems ought to be overcome in such an approach: description of the accompanying intervalence chemical reactions, general scheme of the mechanism, estimation of stability of low-valence intermediate species and revision of some deeply rooted concepts inconsistent with the one-electron representation of the overall process. The first 3 chapters of the book are devoted to these problems.

To begin with, the basic concepts and definitions related to the model of stepwise processes are introduced and rigorously formulated. This model presumes a series of consecutive one-electron electrochemical reactions with conjugated chemical intervalence reactions, every low valence intermediates taking part in these reactions. The properties of a particular system, including the possibility to observe the separate steps, follow logically from the stability of the intermediates that can be characterised by either equilibrium or rate constants of the intervalence reactions.

Provided the solubility of low-valence intermediates (LVI) is low, the stepwise character of the process should result and often does result in the formation of three-phase system "electrolyte-film of LVI-metal"; we call it "electrochemical film system" (EFS). Such situation is very common in a long-term electrolytic process. Thus, the second main aspect of the approach is to consider the film as an active participant of the process rather than simply as a passivative layer. This point of view is also defined and substantiated in the first chapter. As a starting point, some ideas on the solid-phase electrochemical reduction mechanism were borrowed from the electrochemistry of refractory metals in aqueous solutions [5].

It is shown that the Velikanov model [6] of polyfunctional conductor (PFC) is a good approximation for the properties of the LVI film. The electrochemistry of PFC had attracted considerable attention in the 1970s with regard to the practical problem of electrochemical processing of chalcogenide (sulphide) compounds, the components of natural polymetallic ores.

Since the properties of PFC are a key point for understanding the behaviour of electrochemical film systems, this subject is reviewed in Chap. 4 with special emphasis on the recent results related to the electrochemistry of ionic liquids.

The classification of electrode film systems is proposed based on the above ideas, and main qualitative regularities of the electrolytic processes in the film systems of different kind are envisaged in Chap. 4. In particular, the mechanism of formation of cathode deposits is considered. It is shown that the deposition of metal–salt "carrots" or compact metal layers depends on the properties of the cathode film system (prevailing type and ratio of the electronic and ionic conductivity of the film). The nature of crisis phenomena at the electrodes is also analysed (anode effect in fluoride melts, complications at the electrolytic production of Al–Si alloys in industrial-scale electrolytic cells), the mechanisms are elaborated and the means to escape the crises situations are developed.

From the standpoint of thermodynamics, the system electrolyte–film–electrode is open and far from equilibrium state. In this study we use the theoretical approach to the description of such systems created by H. Poincaré and further developed later by Andronov and others. This method is called bifurcation analysis or, alternatively, theory of non-linear dynamic systems [7]. It has been applied to the studies of macrokinetics (dynamics) of the processes in electrode film systems.

There, the mathematical model is represented by a system of non-linear differential equations based on the balances of extensive physical quantities like mass, energy and charge. General rules for construction and analysis of such models are given in Chap. 1, while their applications to the specific systems are discussed in Chap. 5. In particular, a theory of anode refinement of a polyvalent metal in long-term processes is considered. It explains the periodic changes of current efficiency in the refinement of Nb and Ta along with the development of crisis phenomenon in pilot-scale long-term refinement process for metallurgical silicon that did not appear in short-term laboratory experiments. A theoretical model of the anode process at carbon electrodes in the condition of anode effect is also considered. This model allows for predictions of some interesting features of the process. For example, it proves the "supercritical" regimes to be possible for the electrolysis at current density of several orders higher than the initial critical value. This possibility is verified experimentally in KF–KBF_4 molten salt mixtures. Finally, as an additional example of the capabilities of the theoretical methods, Chap. 5 includes the developed mathematical model of self-discharge of highly energy-intensive lithium batteries. It predicts the situations when instability occurs developing finally into thermal runaway and explosion of the cell during storage with no apparent external reasons.

Hence, the purpose of this book is to provide a unified basis for a wide range of problems relevant to the electrochemistry of many-electron processes in ionic melts and in other media as well. Equilibria in many-electron systems, non-stationary many-electron processes, electrochemical processes in mixed conductors, aspects of the electrodeposition of polyvalence elements and anode processes are considered. No arbitrary assumptions like one-step many-electron transfers or "discrete" discharge of complex species are involved—the consideration is based on a few very general ideas.

Of course, the results presented are neither a final solution to all problems nor a completed phenomenological theory of many-electron electrochemical processes; rather, this is an attempt to make a few steps towards such theory. Since the whole problem is challenging and complex, some results are not flawless; we are quite aware of that. Especially this is true for the kinetics of many-electron processes: the proposed mechanism is a sketch rather than a rigorous description. Perhaps, some conclusions are disputable requiring further discussion and elaboration.

Anyway, this book is an attempt to find order in the huge set of separated facts accumulated by numerous works in the electrochemistry of many-electron processes. Pursuing this aim, we were inspired by the ideas of [8]: science has to find an order and simplicity under apparent disorder and complexity of the observed facts. Further on, new complexity is hidden under this simplicity; it is already looking out when we consider the systems far from equilibrium—to find new simplicity under this new complexity is a task for future researchers. And so on, *ad infinitum*.

References

1. Delimarskii YuK, Markov BF (1960) Electrochemistry of molten salts. Metallurgizdat, Moskow (in Russian).
2. Baimakov YuV, Vetiukov MM (1966) Electrolysis of molten salts. Metellurgizdat, Moscow (in Russian).
3. Delimarskii YuK (1978) Electrochemistry of ionic melts. Metallurgia, Moscow (in Russian).
4. Dogonadze RR, Kuznetsov AM (1978) Kinetics of heterogeneous chemical reactions in solutions. In Kinetics and catalysis. VINITI, Moscow, pp 223–254
5. Vas'ko AT, Kovach SK (1983) Electrochemistry of refractory metals. Tekhnika, Kiev (in Russian).
6. Velikanov AA (1974) Electronic–ionic conductivity of non-metal melts. In Ionic melts, vol 2. Naukova Dumka, Kiev, pp 146–155
7. Strogatz SH (1994) Nonlinear dynamics and chaos with applications to physics, biology, and engineering. Cambridge, MA, Westview Press
8. Poincaré H (1982) The foundations of science: science and hypothesis; the value of science; science and method. UPA

Contents

Chapter 1
Many-Electron Electrochemical Systems: Concepts and Definitions

Law < of Nature > is taken to be simple until the opposite is proven.

H. Poincaré

1.1 Definition of a Stepwise Electrochemical Process

Hereby, an electrochemical process is defined as a many-electron one when the oxidation states of a reagent and final product differ by more than 1, thus requiring the transfer of more than one electron to transform, "electrochemically", at a defined current flow, the reagent into the product.

The stepwise description of such process necessarily follows from general principles of chemical kinetics: a third order reaction is unlikely; a reaction of higher order is improbable. If electrochemical reactions are considered, a direct two-electron process should be thought of as a third order and, hence, as an unlikely reaction.

Quantum theory of an elementary electron transfer act confirms this suggestion. In the early 1970s, using Marcus' idea on the fluctuations of solvent energy as a driving force for electron transfer [1], Vorotyntsev and Kuznetsov [2] showed theoretically that, for non-adiabatic reactions, the elementary two-electron step is highly improbable, while Dogonadze and Kuznetsov proved that the steps with more than two transferred electrons are practically impossible [3]. It is consistent with the rules of chemical kinetics mentioned above: two-electron elementary step can formally be presented as almost improbable reaction of third order, and three or more electron steps as the impossible reactions of more than third order.

Lately, Karasevskii and Karnaukhov [4] considered the adiabatic many-electron reactions. They showed that, at a certain (adiabatic) conditions, when the system is far enough from equilibrium state, the energy minima corresponding to the

A.A. Andriiko et al., *Many-electron Electrochemical Processes*,
Monographs in Electrochemistry, DOI 10.1007/978-3-642-35770-1_1,

intermediate oxidation states disappear; thus, several electrons can follow to the final state from/onto the electrode practically simultaneously. Thus, for electrochemical states with wave functions strongly overlapping (adiabatic behaviour) the possibility exists for apparent one-step transfer of at least two electrons. Perhaps, this mechanism is similar to the synchronous "concerted" electron transfer proposed for homogeneous redox systems [5]. Again, when the adiabatic many-electron system is shifting nearer to equilibrium (by lowering the current density), the intermediate energy minima reappear, thus transforming the apparent process to a sequence of one-electron steps.

Unfortunately, the above theories of elementary electrochemical reactions cause very little impact on the experimental research practice. It is still widely accepted that a current–potential curve of a many-electron reaction, if not split into several waves, can be evaluated assuming a one-stage many-electron transfer on the basis of simple equations derived for a one-step, one-electron reaction. Usually, it looks as follows. If only one electrochemical wave in the current–potential curve is found corresponding to the overall n-electron process, then the conclusion is drawn that the process proceeds through one n-electron step. The curve is analysed in terms of one of the electrochemical equations derived for typical one-electron process, some coefficient $RT/\alpha n_a F$ is calculated, where α means the "transfer coefficient" of the one-step reaction with n_a electrons being transferred together. When the value of $\boldsymbol{\alpha}$ is close to 1, the process is considered as reversible (Nernstian), otherwise as irreversible (or quasi-reversible) n-electron step. As an example, see [6, 7] and references therein concerning the reduction of WO_4^{2-} in molten halides.

The above situation is common in electrochemistry of molten salts, especially in the description of processes related to the production of refractory metals. A large number of reaction mechanisms were proposed including one or more one-stage many-electron reaction: $W(VI) + 6e^- \rightarrow W$ and $Mo(VI) + 6e^- \rightarrow Mo$ [6–8]; $Ta(V) + 5e^- \rightarrow Ta$ [9]; $Nb(IV) + 4e^- \rightarrow Nb$ [10, 11]; $U(III) + 3e^- \rightarrow U(0)$ [12]; and many other similar schemes are currently in use. Obviously, such approach is ambiguous, is sometimes meaningless (for example, what is the physical meaning of the transfer coefficient $\boldsymbol{\alpha}$ for a number of electrons more than 1?), and has a little prediction capability. Why then such descriptions are so widespread? One of the reasons is that almost each of electrochemical handbooks, both on molten salts [13] and general [14], offers these "$\boldsymbol{\alpha n_a}$"—equations for irreversible electrochemical reactions without even mentioning the fact that all such equations were derived for a simple one-step one-electron reaction. To our knowledge, only the second edition of Bard and Faulkner book [15] is an exception.[1]

[1] Curious footnote on page 109 of [15] says:

"In the first edition and in much of the literature, one finds n_a used as the n value of the rate-determining step. As a consequence n_a appears in many kinetic expressions. Since n_a is probably always 1, it is a redundant symbol and has been dropped in this edition. The current–potential characteristic for a multistep process has often been expressed as <Butler–Volmer equation with αn_a coefficient>. This is rarely, if ever, an accurate form of the i–E characteristic for multistep mechanisms."

At last!

The opposite approach considering a one-electron multistep mechanism can be found in many research publications related to the electrochemistry of aqueous solutions. It originates from early works of Vetter [16], where a many-electron process is represented as a sequence of one-electron steps at the electrode surface. Rather cumbersome equations were obtained for the stationary current–potential curves, transforming into the known electrochemical equations in simplest limiting cases [15–18]. The weakness of this model consists in representation of the intermediate products as merely surface states which are unable to leave the electrode in course of the process. The experience shows that such assumption is physically unreal. Hence, the model is restricted only to some specific adsorption processes.

In this book we will follow the concept of one-electron stepwise discharge that differs from the Vetter's one. It can be formulated in the following terms:

1. The assumption on the one-step many-electron transfer is not used. Any elementary electrochemical reaction of many-electron process is represented as a one-electron step.
2. Every possible low valence intermediate (LVI) is formed in such a process.
3. Every LVI enters chemical redox interactions in the bulk. According to [19], we define these interactions as *intervalence reactions* (IVR).
4. The concept of IVR comprises all reactions of electron exchange between the species of polyvalent element in every oxidation state. The disproportionation reaction, when two identical species with equal oxidation state react, is a particular case of IVR.
5. An elementary IVR is taken to be also one-electron. That is, two particles can exchange no more than one electron in one elementary action. For example, the reactions

$$2\mathrm{Si(II)} \leftrightarrows \mathrm{Si(III)} + \mathrm{Si(I)} \tag{1.1}$$

or

$$\mathrm{Si(IV)} + \mathrm{Si(I)} \leftrightarrows \mathrm{Si(III)} + \mathrm{Si(II)} \tag{1.2}$$

are elementary LVR, but

$$2\mathrm{Si(II)} \leftrightarrows \mathrm{Si(IV)} + \mathrm{Si(0)} \tag{1.3}$$

should be considered only as a notion for an overall scheme of a complex multistep process.
6. *Stability* of a LVI is an important parameter. It can be defined in either thermodynamic or kinetic terms. Equilibrium constants of IVR determine the thermodynamic stability, and rate constants determine the kinetic stability of LVI. A fixed set of independent equilibria should be used for quantitative description of thermodynamic stability. This will be discussed in the next chapter.

7. The apparent character of many-electron processes, and thus the experimental possibility to observe the one-electron electrochemical steps, depends on the stabilities of the LVIs, either thermodynamically or kinetically defined. A process should appear as one-stage many-electron one at low stabilities of LVIs and will split into sequential reactions at higher stabilities. The intermediate region exists where the overall process is represented by one distorted electrochemical wave in an electrochemical curve. Thus, the discussion on one- or many-electron electrochemical steps is becoming of quantitative rather than qualitative nature.
8. Low solubility of a LVI could invoke the formation of new solid phases at the electrodes. The number of possible phases at the conditions not far from equilibrium ones is determined by Gibbs' rule. At constant temperature and pressure, it states:

$$S_{\max} + 1 = \kappa \tag{1.4}$$

where $S_{\max}$ is a maximal number of separate phases of the compounds in all oxidation states and κ is a number of components.

For a two-component system, for example, $Pb/PbCl_2$, $S_{\max} = 1$, that is, in the presence of the individual metal phase, the LVI cannot form a separate thermodynamically stable phase. For a three-component system, e.g., Si/K_2SiF_6–KF, co-existence of solid Si and one solid phase of an intermediate is possible; if SiO_2 is added to the melt, then two stable solid intermediate phases are thermodynamically possible, etc.

No basic ideas other than mentioned above are used. Thus, the developed approach is not used as a microscopic insight; in other words, it is a purely phenomenological concept. Hence, it is rather general being not restricted by any model assumptions.

1.2 Stepwise Many-Electron Process: Problem of Kinetic Description

A rigorous description of the kinetics of a many-electron process should, in principle, be based on the solution of the system of differential equations of reactive diffusion written down for all soluble species including intermediate low valence compounds (LVI):

$$\frac{\partial Y_i}{\partial t} = D\nabla^2 Y_i + W_i(Y) \tag{1.5}$$

where Y_i is the concentration of component Y_i, D is the diffusion coefficient, and ∇^2 is the Laplace operator. The kinetic term $W_i(Y)$ denotes the concentration dependence of the rate of accompanying chemical reactions. It is determined by all accompanying elementary IVRs that, according to clause 5 of the section above, can be denoted as follows:

$$E(i) + E(j) \leftrightarrows E(i \pm 1) + E(j \pm 1) \tag{1.6}$$

where oxidation states of the species are given in brackets; $0 \leq i, j \leq N$, N being the maximal possible oxidation number.

In the attempt to develop such description, two complications arise:

Firstly, the accompanying reactions (1.6) are of second order; thus, the kinetic terms $W_i(Y)$ are quadratic and a system of differential equations (1.5) becomes non-linear.

Secondly, it is easy to show that, if the highest oxidation number is N, the total number of the possible reaction of Eq. (1.6) type, including the reverse ones, is equal to

$$q = N(N - 1). \tag{1.7}$$

Evidently, it becomes practically impossible to take into account this large number of reactions when $N > 2$ (for example, $q = 12$ for four-electron process like reduction of Ti(IV) to Ti metal).

Moreover, a superposition of other chemical reactions (like the dissociation of components, interaction with supporting electrolytes, and others) is also possible in a real electrochemical process.

Thus, a discouraging conclusion follows that an absolutely accurate description is practically impossible and some simplifying assumptions are inevitable.

A simplified kinetic scheme of a many-electron process has been proposed which takes into consideration not every but only those of the reactions (1.6) where $i = j$, that is, $N - 1$ disproportionation reactions of general kind [16, 20–22]

$$2E(i) \leftrightarrows E(i + 1) + E(i - 1) \tag{1.8}$$

Then, for example, the general reaction scheme of four-electron process can be written as follows:

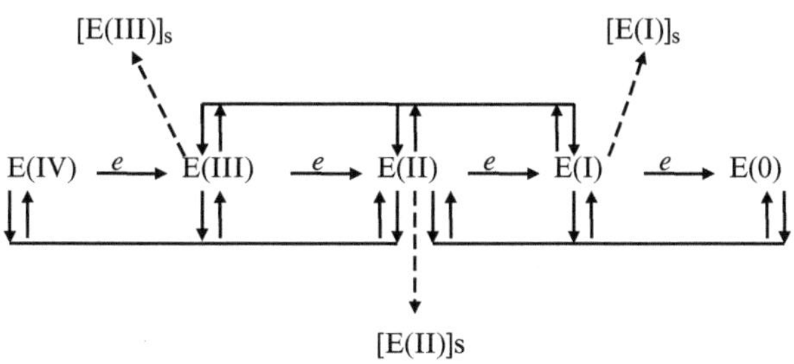

(1.9)

Here the contour I denotes the reaction (1.8) where $i = 1$, and the II and III are for $i = 2$ and 3 accordingly. Solid intermediate species are denoted by index s. The arrows $\underset{\rightarrow}{e}$ represent heterogeneous reactions of electron exchange between particles and the electrode surface. The dashed arrows point to the possibility of deposition of solid LVIs on the electrode, resulting in the formation of an *electrode film system* (see below).

Sometimes, other reactions [e.g., previous dissociation with formation of the electroactive species E(IV)] have also to be considered.

Application of the approximate scheme (1.9) to the description and interpretation of the experimental studies of many-electron processes is given below in Chap. 3.

1.3 Electrode Film Systems. Formation, General Properties, Classification

Supposing the LVIs of a many-electron process have low solubility in the supporting electrolyte, they can form a deposit on the electrode surface in course of discharge [see the scheme (1.9) above] causing the formation of a three-phase system electrolyte–film–metal. After Vas'ko [23], we shall call this structure an *electrode film system* (EFS).

A central and most important idea of EFS is to consider the film not as a simple passive layer but as an active participant of the process. Particularly, in a cathodic process, it can be electrochemically reduced directly to the metal by solid-state electrochemical mechanism.

Upon our knowledge, Gerischer and Käppel [20] were the first who mentioned this possibility with regard to the electrochemical deposition of chromium from aqueous electrolytes containing Cr(VI) compounds. It was then only a guess; the EFS was first cast explicitly in works of Vas'ko [21, 23], who made the first attempt to correlate the behaviour of the electrode system to electrophysical properties of the film. Later on, similar ideas were also used by Gorodyskiy and co-workers in a series of papers on the electrodeposition of some metals from fluoride-containing aqueous electrolytes [see review [22] and references therein]. These authors use the notation "bifunctional electrode system" in a sense that the film possesses a double function, acting as an electrode with regard to the outer liquid electrolyte and as a solid electrolyte relatively to inner metal electrode.

These ideas, which come from electrochemistry of aqueous solutions, have been further developed and applied to the processes in molten salts media [24, 25]. It was found that the properties of the intermediate film could be represented in terms of the concept of a polyfunctional conductor (PFC), which was invented and described by Velikanov [26]. This phenomenological concept is based on the theory of amorphous [27] and liquid [28–30] semiconductors. Since it is important for our purposes, it will be discussed in more detail below in Chap. 4. Here we should only emphasise a few main features of the object:

- The PFC has electronic and ionic electric conductivity simultaneously.
- The electronic constituent can be of metallic or semiconductive nature.
- The PFC is considered to be amorphous, that is, in case of semiconductive nature of electronic conductivity, it cannot be doped retaining its intrinsic conductivity.

Hence, our model of EFS is represented by consecutively connecting a conductor of 1st kind (electronic)—the PFC—and a conductor of second kind (ionic). The description of the EFS is thus focused on the concept of polyfunctional conductors. For this reason, we use the earlier introduced term "film system" as more appropriate than the "bifunctional" one.

1.4 Velikanov's Polyfunctional Conductor

The concept was elaborated as the expansion of the theory of liquid semiconductors [30] applied to the problem of electrochemical decomposition of liquid chalcogenide compounds [26, 31]. In particular, basic components of natural metal ores, non-oxidised metal sulphides, exhibit the properties of PFC in the fused state. According to Velikanov, the PFC is an idealised complex ion–electron conductor where several electron transport mechanisms (ionic, semiconductor, metallic) of different physical nature contribute to the overall conductivity, and each of these contributions exhibits its own temperature dependency of the conductivity [26, 31].

Thus, the ionic contribution is weakly dependent on temperature increasing approximately linearly as the temperature increases.

Remarkably, the known Arrhenius-type equation

$$\sigma = \sigma_0 \cdot \mathrm{e}^{-\frac{\Delta E}{2kT}} \tag{1.10}$$

derived for solid semiconductor remains valid for the semiconductor part of conductivity of PFC; hence, this contribution increases exponentially with temperature. As the semiconductor degenerates at higher temperatures, the electronic part of conductivity becomes metallic. It decreases slowly as the temperature increases. Then the analysis of the experimental temperature dependence (polynomial) of conductivity allows determining the dominating charge transport mechanism and calculating the separate contributions to the overall conductivity. From this, one can derive the conditions for the electrochemical decomposition of the PFC, which is possible when ionic contribution is large compared to the electronic one.[2]

[2] Within this approach, a number of research works were carried out in Kiev in 1960–1980s. The ways to affect the nature of ionic-electronic transport were studied. As a result, direct electrolytic methods were developed for production of some nonferrous metals based on the phenomenon of suppressing electronic conductivity with some "heteropolar" additives like alkali metal sulphides, chlorides, and some other compounds [32–36].

The electrochemical behaviour of a PFC is of prime interest for our purposes. When separating the total current into electronic and ionic parts, it is usually assumed that the ratio of these particular currents is equal to the ratio of electronic and ionic conductivities. We shall use this simple model of parallel connection of electronic and ionic resistors for qualitative classification of EFSs (see below). However, as it has been shown more recently, such model is only a rough approximation—the ratio does not remain constant with change of the applied current because of non-linear dependence of the partial conductivities on current [37, 38]. This problem will be discussed in Chap. 4.

1.5 Classification of Film Systems. Processes in Chemical Film Systems

The qualitative classification of film systems, both chemical and electrochemical, was proposed in [25]. It is based on the prevailing type and magnitude of conductivity of the film considering the film as a PFC as mentioned above. It is as follows:

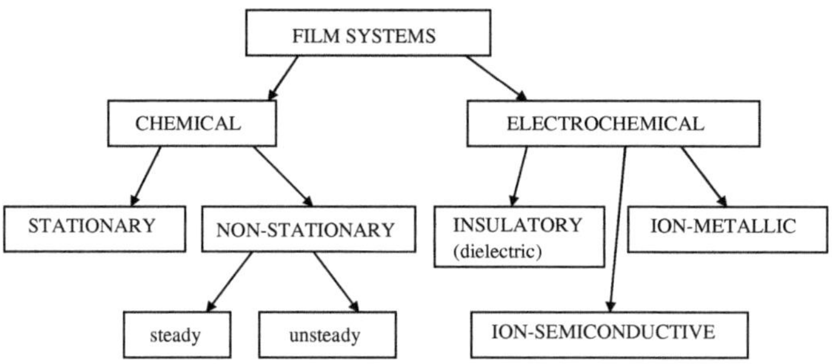

(1.11)

The electrolysis process is essentially different for each type of EFS. These peculiarities will be considered in Chap. 4.

One can see that the classification scheme (1.11) points out the separate group of chemical film systems where the process does not involve electric current across the interface. Let us briefly overlook the main features of such systems.

1.5.1 Processes in Chemical Film Systems

If a solid substance A_s reacts with a dissolved component U forming a poorly soluble compound V, then the latter can precipitate at the surface forming the film. This can be represented by the approximate scheme

$$
\begin{array}{l}
A_s + U \xrightarrow{1} V \xrightarrow{2} S \\
\qquad\qquad\quad V \xrightarrow{3} V_1
\end{array}
\qquad (1.12)
$$

The product V of reaction 1 forms the surface deposit S (reaction 2) and partly is transferred into the bulk of liquid by diffusion. The component V can be as well the low soluble intermediate entering further into reaction 3 with components of the solvent (melt) and forming more soluble substance V_1. Both of these cases result in the formation of chemical film system.

A number of such systems were found experimentally in the reactions of ceramic materials with ionic melts, for example, quarts with molten alkali [39], alumina with molten borax [40], ceramic oxides (Al_2O_3, SiO_2, ZrO_2, porcelain) with fluoride melts [41], germanium with oxyfluoride melts [24], and aluminium with cryolite–alumina melts [42, 43].[3]

Figure 1.1 represents the schematic drawing of a chemical film system resulting from the reactions (1.12).

Symbols $\boldsymbol{\lambda}$ and $\boldsymbol{l}$ in Fig. 1.1 correspond to the thickness of film *S* and solubility of the component V in the bulk, and dashed lines indicate the tentative non-stationary distribution of concentrations of species V and U after the deposit is formed.

The stationary thickness of the film λ_{st} is expressed by the formula:

$$
\lambda_{st} = \alpha\delta\left(\frac{U_0}{l} - 1\right) - \frac{D}{k} \qquad (1.13)
$$

where δ is the thickness of diffusion layer in the liquid, U_0 is the concentration of the component U in the bulk, l is the solubility of the deposit (component V), D is the diffusion coefficient, k is the rate constant of heterogeneous reaction 1 of the scheme (1.12), and α is a parameter accounting for the permeability of the deposit S for the diffusion flux of component U to the boundary ($0 < \alpha < 1$).

Analysis of the relation (1.13) reveals three possible specific types of chemical film systems:

1. If $0 < \lambda_{st} < L$, where L is a characteristic dimension of the whole system (for example, a distance from the sample to the wall of the experimental cell, or diameter of the vessel), then a *stationary* system is finally formed with finite thickness of the film (type 1). Intuitively, the stationary state seems to be stable, although this problem was not studied theoretically.

---

[3] The two last examples represent the FS intermediate between chemical and electrochemical ones, because the interaction follows the mechanism of electrochemical corrosion: overall current is zero, though partial currents flow with opposite sign, typical for corrosion situation. It seems that this kind of systems comprises also the processes of metal oxidation which obey Wagner's theory (see Chap. 4).

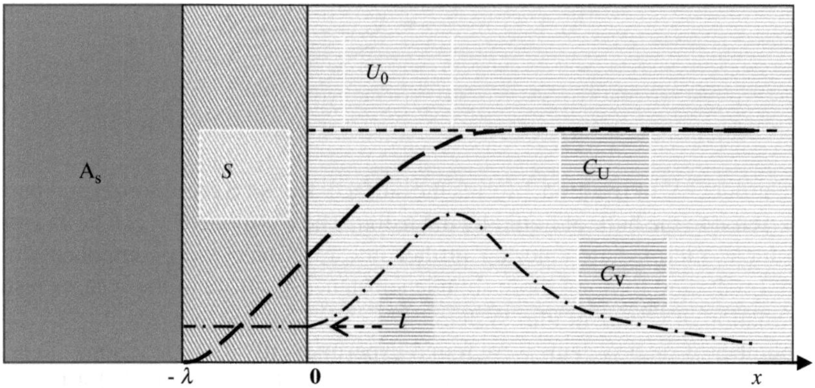

Fig. 1.1 Schematic presentation of a chemical film system. Approximate concentration profiles are shown of the reagent U and product V the solubility of which is l

2. If $\lambda_{st} > L$, which could happen at very low solubility l, the stationary state, however being possible theoretically, becomes unattainable practically. The behaviour of such *non-stationary steady* system is governed by the well-known Tamman law for parabolic growth of the film (type 2).
3. Finally, if Eq. (1.13) gives $\lambda_{st} < 0$, it does not necessarily mean the absence of the film in the system. The situation is also possible (e.g., very small α or k) when the film is formed; however, its positive stationary thickness is impossible. These *non-stationary unsteady* systems can exist either at partial covering of the surface or in the regime of periodic formation and dissolution of the film (type 3). The behaviour of such systems is somewhat similar to the unsteady electrochemical systems (see Chap. 5).

In general, a theoretical description of non-stationary (transient) processes in any chemical FS (especially, of third type) is a very difficult mathematical problem. It consists in solution of system of differential equations for diffusion of the components in moving coordinate system shown in Fig. 1.1. These equations are highly non-linear:

$$\frac{\partial V}{\partial t} = D\frac{\partial^2 V}{\partial x^2} + \dot{\lambda}\frac{\partial V}{\partial x} \tag{1.14}$$

(similarly, for component U). Here, $\dot{\lambda} = \mathrm{d}\lambda/\mathrm{d}t$ is the rate of film's grows that is determined by the values of fluxes at the boundaries. Hence, the coefficient at the first derivative with respect to the coordinate contains implicitly the boundary conditions in terms of the first coordinate derivatives at the interfaces of the system.

Presently, this problem has defied solution. It could only be mentioned that, at $l \approx 0$ and $\alpha \neq 0$, the asymptotic solution is the Tamman's parabolic law, as for the case 2 above. Though, this result, reflecting the proportionality between the values

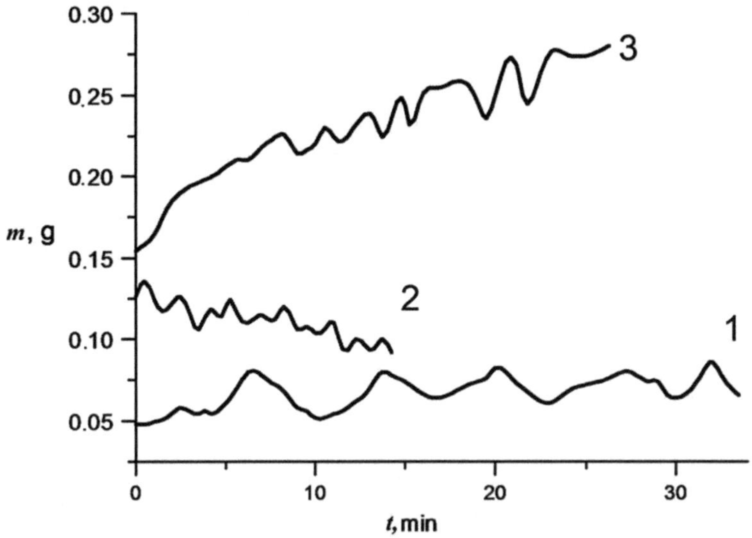

Fig. 1.2 Oscillations of weight in some non-stationary chemical film systems: (1) Alumina sample in cryolite melt with SiO_2 additives; (2) germanium metal in the molten mixture KF–NaF–K_2GeF_6–GeO_2; and (3) fresh cathode deposit of germanium in the melt as in the case 2

of coordinate and square root of time, is trivial for diffusion problems that do not include any data with the dimension of time or length in boundary conditions [44, 45] and, thus, is of not much interest. As a matter of fact, the behaviour of real systems is much more complex and variable. For example, macroscopic oscillations of the weight of sample mass have sometimes been observed. Figure 1.2 shows some examples of such processes [24, 25] registered with the experimental method of automatic balance described in [46].

These considerations point to the qualitatively unusual nature of film systems comparatively with common subjects of electrochemistry. Even in the simplest case not involving the current passed through the interface, we deal with systems where traditional methods of chemical thermodynamics and kinetics are helpless. Nevertheless, the development of mathematic models is sometimes possible to account for main regularities of macrokinetics[4] in such systems. The basic ideas and methods of this modelling will be given below in the final section of this chapter.

[4] Under this term we understand the regularities of the processes developing in the system in time on long-term scale, from tens minutes to several days or more, which results in relatively slow changes of macroscopic variables. A similar term is "dynamics"; it is used more often in physical papers and will be used in this book as well.

1.6 Macrokinetics of Processes in Film Systems: General Principles of Theoretical Modelling

From the thermodynamic point of view, the systems under consideration are open and far from equilibrium state. Obviously, common physico-chemical methods of research and theoretical description fail to be helpful in this case.

Up to date, significant progress in the development of theoretical methods for such non-equilibrium systems has been achieved. Three main approaches can be outlined as follows:

1. Theory of dissipative structures.
2. Statistic theory of macroscopic open systems.
3. Catastrophe theory.

Eventually, all of them are based on the methods of general qualitative theory of differential equations developed by Poincaré more than a century ago [47]. This theory was essentially developed by Andronov in 1930s [48] and, finally, after Hopf's theorem on bifurcation appeared in 1942 [49]; it became a self-consistent branch of mathematics. This subject is currently known under several names: Poincaré–Andronov's general theory of dynamic systems; theory of non-linear systems; theory of bifurcation in dynamic systems. Although the first notion is, in our opinion, the most exact one, we will use the term "bifurcation theory", or BT, for the sake of brevity.

Theory of dissipative structures (or theory of self-organisation) was born as a result of fusion of BT and ideas of irreversible thermodynamics developed mainly by the "Brussels school" of physicists (Prigogine and co-workers) [50–52]. This approach, like a masterpiece of art, reveals the harmony of Nature giving the unique physical background for a widest range of phenomena, from hydrodynamics to life; thus, its philosophic value is beyond any doubt. However, one can agree also with Murrey [53] who suggests that the formalism of self-organisation theory provides nothing new and useful for any specific problem, comparatively with pure methods of BT. Really, it seems very likely that the theory of dissipative structures is a presentation of the theory of dynamic systems in terms of irreversible thermodynamic language rather than a self-contained theory.

Catastrophe theory is based, again, on the BT methods and Whitney's theorem on the singularities of smooth manifolds [54] related to such abstract science as mathematics topology. Emerging in the 1960s [55], it had soon gained wide popularity due to philosophic works of the founder, French mathematician Thom [56, 57]. It was tried to be applied to almost any branch of knowledge, from mechanics to economy and policy. However, some scientists consider this theory as imperfect and speculative [55, 58]. Anyway, from the very beginning and by virtue of its generality, the catastrophe theory was intended only for qualitative understanding how the old structures (symmetries) are lost and new ones are born [57]. Consequently, it is of little use for the development of quantitative (or even semi-quantitative) models of physico-chemical systems.

Statistic theory of macroscopic open systems represents an attempt to combine the dynamic (Poincaré–Andronov) and statistic (originated from Boltzman) approaches to open non-equilibrium systems [59]. This theory looks promising since it gives rise for more complete description of the systems. In particular, it could, in principle, predict the behaviour of the system near the bifurcation point that mere BT is not able for. However, this modern sophisticated theory is still in a state of development and far from been completed. In our opinion, it has a little chance to be easily applied to physico-chemical systems at present.

Following conclusions can be drawn from the brief review given above.

The statistical theory of open systems is not yet developed enough to be applied to physico-chemical problems. Both catastrophe and dissipative structure theories are of more general philosophic rather than practical value. So, only the classic Poincaré–Andronov's bifurcation theory gives real tools for the formulation and investigation of the mathematical models of the processes developing in physical and chemical systems far away from equilibrium. Some examples are presented in Chap. 5 where these tools were successfully applied to electrochemical systems. Main principles of such applications are given below.

1.6.1 Mathematical Models of Physico-chemical Processes Using an Approach Based on the Bifurcation Theory

Before mathematical modelling, an appropriate physical pattern must be envisaged on the basis of experimental data. Since the nature of real systems is complicated, the choice of the physical model is ambiguous in most of the real situations. Several possibilities exist that cannot be preferred beforehand. The adequacy of the physical model has then to be estimated qualitatively. Preliminary it can be done considering the character of feedbacks in each possible model.

Under the *feedback* in dynamic system we understand the effect of variables on the rate of change of the variables. If the rate increases as a variable increases, the feedback is defined as *positive*; otherwise it is *negative*. A feedback is *intrinsic* when a variable affects its own rate of change and *crossed* if one variable affects the rate of other variable's change.

The idea of *stability* is a central point of BT. Some qualitative information on the stability of system's stationary state often follows immediately from the character (or direction in an above sense) of the feedbacks, before any mathematical model is formulated.

In many cases the physical model can be reduced to two-dimensional dynamic system, that is, with two independent variables. For instance, in a group of thermokinetic models, the behaviour of system is governed by equations of energy and material balances, and thus, content (concentration) of a component and temperature could be the independent variables. Following rules are valid for such kind of systems:

If intrinsic feedbacks are negative and crossed ones are of different sign, all stationary points of the system are stable. If all intrinsic feedbacks are positive, no stable stationary state exists in such system. Both stable and unstable stationary states of the system are possible at any other combination of the feedbacks.

The situation may occur when stationary state of the system sharply transforms into another one if a value of certain parameter passes through some critical point. The character of the new state after this event (which is called *bifurcation*) depends on the type of *attractor* in the phase space of the system in the vicinity of the old steady state that lost its stability after the bifurcation. The idea of attractor plays an important part in BT and means "attractive manifold" that pulls together all tracks (locuses) of a system at infinite (in time) distance away from the initial state.

Two types of attractors were known since the times of Poincaré: points or closed curves (limit cycles). The third type was discovered in 1971 [60]. It is so-called "*strange attractor*", which can exist in three- and more-dimensional systems. In accordance with these three known types of attractors, three different kind of system's behaviour are possible after bifurcation: (1) transition into a new stable steady state; (2) undamped self-oscillations, and (3) chaotic regime (turbulence).

Bifurcations with self-oscillation scenario are most probable in the systems where the pairs of both intrinsic and crossed feedbacks have opposite signs. If the signs of the crossed feedbacks are the same, the oscillations are less probable.

After these preliminary qualitative physical considerations, the development of mathematical models has to be the next step of the study.

Such a mathematical model of a non-equilibrium physico-chemical system is represented in terms of a system of differential equations (as a rule, essentially non-linear). These equations are derived from appropriate physical balances of substances, energy, and charge. Following rules are useful in the process of construction of the model:

- The balance differential equations are written in form of dependencies of the rates of change (time derivatives) of system's natural variables (like concentration, temperature, etc.) on the values of these variables.
- Then, by an appropriate choice of scaling for the variables, the differential equation system is brought to a dimensionless form where time derivatives of dimensionless variables are the functions of these new dimensionless variables and dimensionless parameters.
- A probable order of magnitude is estimated for each dimensionless parameter. Judging from these values, the parameters are chosen that could be neglected and then all possible simplifications of the model are made.

As a result, a mathematical model is obtained in form of a system of differential equations:

$$\dot{X}_i = f_i(X, \alpha_k) \tag{1.15}$$

where X are the variables, α the parameters, f the (non-linear) functions of the variables and parameters, and index $i = 1, \ldots, n$, where n is the n-dimensionality of the system.

Mathematical investigation of the model (1.15) includes the following steps:

1. Determination of stationary states (points of rest) of the system.
For this, the system of algebraic equations

$$f_i(X_{\mathrm{st}}, \alpha_k) = 0 \tag{1.16}$$

has to be solved.
2. Analysis of stability of stationary states.
By Lyapunov, the solution of system is asymptotically stable if, for an arbitrary however small value ε, such value of time τ exists that $|X - X_{\mathrm{st}}| < \varepsilon$ at $t > \tau$.
The solution of non-linear system (1.15) is stable if the solution of the linearised system

$$\delta \dot{X}_j = \Sigma \frac{\partial f_i}{\partial X_j} \delta X_j \tag{1.17}$$

where $\delta X_j = X_j - X_{j_{\mathrm{st}}}$ denotes the deviation from equilibrium state, is also stable.
A partial derivative in Eq. (1.17) is a mathematical definition of the system's feedback (see above): the feedback is intrinsic if $i = j$ and crossed if $i \neq j$.
Analysing the matrix of coefficients of the system (1.17)

$$A = \left(\frac{\partial f_i}{\partial X_j} \right) \tag{1.18}$$

one can determine the stability of the stationary points of the system (1.15).
The condition of stability is the following:

$$Sp(A) < 0; \quad J(A) > 0 \tag{1.19}$$

where *Sp*(*A*) is the sum of diagonal numbers, or "*spur*", of the matrix *A* (that is, the sum of derivatives $\partial f_i / \partial X_j$ at $i = j$), and $J(A) = \det A$ is the determinant of matrix (1.17), or "*Jacobian*" of the system (1.17).
The inequalities (1.19) are the mathematical notation for qualitative statements above. In particular, they are met immediately at all intrinsic feedbacks negative and crossed ones of opposite sign; the stability may be lost once a positive intrinsic feedback occurs.
3. Determination of bifurcation points.
A bifurcation (loss of stability of a stationary state) takes place as some critical set of parameters α_k of Eq. (1.15) is achieved when the inequalities (1.19) become broken. Hence, the values of parameters α_k at bifurcation points are determined by the relations (1.19) written as the equations. Usually, the analysis results in the construction of the plot (diagram) dividing the whole domain of parameters into "stable" and "unstable" parts. Having known the orders of values of the parameters of a real system, one can use such "bifurcation

diagram" to judge about the possibility of bifurcation in the system. If the bifurcation(s) is (are) possible, then the next step is:

4. Investigation of the system's behaviour after the bifurcation.
   If variation of some parameters (current, temperature, etc.) brings about the loss of stability, following possibilities exist after the bifurcation:

   (a) *Jump into another stationary state*. This new state could be physically unreal (for example, the temperature rising up to infinity in some thermokinetic models, the so-called "thermal runaway"). Then the behaviour of the system would be really "catastrophic"; for instance, arc discharge at the electrode in "anode effect" regime, or self-destruction of a chemical power source (see Chap. 5).
   (b) *Origination of undamped self-oscillations*. This version of the post-bifurcation event (so-called Hopf's bifurcation [49]) was studied most widely, first, relatively to non-linear electric oscillations [48, 61], and, later, to chemical [62, 63] and biological [53, 64, 65] oscillators.
       A positive value of the Jacobian (1.19) in the bifurcation point is one of the conditions for oscillatory regime. This is an equivalent of the assertion that the self-oscillations are most probable in the systems where crossed feedbacks have opposite signs. There are some other methods to identify the self-oscillatory systems: direct application of Hopf theorem, analysis of type of singular points, Bendixson criterion, reduction of the equation system to Lienard equation, and others. One can find details, for example, in Chap. 4 of [53], or elsewhere [65].
   (c) *Transition to the regime of dynamic chaos*. This kind of behaviour is caused by existence of "strange" attractors [60]. Rössler [66] first discovered this regime in physico-chemical systems in 1976; presently, this problem is still poorly studied.

Establishing the bifurcation point and finding out the post-bifurcation trends are the main and most valuable results of the theoretical study of non-equilibrium physico-chemical systems. After that, the investigation of mathematical model is essentially completed. Sometimes, in simplest cases, one can derive an approximate analytical solution for evolution of the system in time, or obtain the numerical solution by computer simulations, and compare these with real experimental dependencies. Though, such solutions could never be rigorous. Because of complexity of real processes and unavoidable simplifications adopted in the model, they would reflect only qualitative trends and, thus, could not be used for quantitative calculations.

Above we have considered the general approaches and methods of non-linear analysis that have been used for studies of macrokinetic models of film systems. Some results are presented in Chap. 5.

References

1. Marcus R (1993) *Nobel lectures* "Electron transfer reactions in chemistry: theory and experiment". Angew Chem 32:8–22
2. Vorotyntsev SA, Kuznetsov AM (1970) Elektrokhimiya 6:208–211
3. Dogonadze RR, Kuznetsov AM (1978) Kinetics of heterogeneous chemical reactions in solutions. In: Kinetics and catalysis. VINITI, Moscow, pp 223–254
4. Karasevskii AI, Karnaukhov IN (1993) J Electroanal Chem 348:49–54
5. Cannon RD (1980) Electron transfer reactions. Butterworth, London
6. Shapoval VI, Malyshev VV, Novoselova IA, Kushkhov KB, Solovjov VV (1994) Ukrainian Chem J 60:483–487
7. Shapoval VI, Malyshev VV, Novoselova IA, Kushkhov KB (1995) Uspekhi Khimii 64:133–154
8. Baraboshkin AN, Buchin VP (1984) Elektrokhimiya 20:579–583
9. Taxil P, Mahene J (1987) J Appl Electrochem 17:261–265
10. Taxil P, Serano K, Chamelot P, Boiko O (1999) Electrocrystallisation of metals in molten halides. In: Proceedings of the international George Papatheodorou symposium, Patras, 17–18 September, pp 43–47
11. Polyakova LP, Taxil P, Polyakov EG (2003) J Alloys Compd 359:244–250
12. Hamel C, Chamelot P, Laplace A, Walle E, Dugne O, Taxil P (2007) Electrochim Acta 52:3995–3999
13. Delimarskii YK, Tumanova NK, Shilina GV, Barchuk LP (1978) Polarography of ionic melts. Naukova Dumka, Kiev (in Russian)
14. Galus Z (1994) Fundamentals of electrochemical analysis, 2nd edn. Wiley, New York, NY
15. Bard AA, Faulkner LR (2001) Electrochemical methods. Fundamentals and applications, 2nd edn. Wiley, New York, NY
16. Vetter K (1961) Electrochemical kinetics. Springer, Berlin
17. Rotinian AA, Tikhonov KI, Shoshina IA (1981) Theoretical electrochemistry. Khimia, Leningrad (in Russian)
18. Sukhotin AM (ed) (1981) Handbook on electrochemistry. Khimia, Leningrad (in Russian)
19. Baimakov YV, Vetiukov MM (1966) Electrolysis of molten salts. Metellurgizdat, Moscow (in Russian)
20. Gerischer H, Käppel M (1956) Z Phys Chem 8:258–264
21. Vas'ko AT, Kovach SK (1983) Electrochemistry of refractory metals. Tekhnika, Kiev
22. Ivanova ND, Ivanov SV (1993) Russian Chem Rev 62:907–921
23. Vas'ko AT (1977) Electrochemistry of molybdenum and tungsten. Naukova Dumka, Kiev (in Russian)
24. Andriiko AA, Tchernov RV (1983) Electrodeposition of powdered germanium from oxide–fluoride melts. In Physical chemistry of ionic melts and solid electrolytes. Naukova Dumka, Kiev, pp 46–60
25. Andriiko AA (1987) Film electrochemical systems in ionic melts. In: Ionic melts and solid electrolytes, vol 2. Naukova Dumka, Kiev, pp 12–38
26. Velikanov AA (1974) Electronic-ionic conductivity of non-metal melts. In: Ionic melts, vol 2. Naukova Dumka, Kiev, pp 146–154
27. Mott NF, Davis EA (1979) Electronic processes in non-crystalline materials. Clarendon, Oxford
28. Ioffe AF (1957) Physics of semiconductors. USSR Academy of Sciences, Moscow-Leningrad (in Russian)
29. Regel AP (1980) Physical properties of electronic melts. Nauka, Moscow (in Russian)
30. Glazov VM, Chizhevskaya SN, Glagoleva NN (1967) Liquid semiconductors. Nauka, Moscow (in Russian)

31. Velikanov AA (1971) Electrochemical investigation of chalcogenide melts. Dr of Sciences Thesis. Institute of General and Inorganic Chemistry, Kiev
32. Velikanov AA (1974) Electrochemistry and Melts (in Russian). Nauka, Moscow
33. Belous AN (1978) Studies on the electrode processes at the electrodeposition of thallium, lead and tin from sulphide and sulphide–chloride melts. PhD (Kandidate) Thesis, Institute of General and Inorganic Chemistry, Kiev
34. Lysin VI (1985) Studies on the nature of conductivity and electrochemical polarization in Chalcogenide–halogenide melts. PhD (Kandidate) Thesis, Institute of General and Inorganic Chemistry, Kiev
35. Zagorovskii GM (1979) Electrochemical investigations of molten systems based on antimony sulphide. PhD (Kandidate) Thesis, Institute of General and Inorganic Chemistry, Kiev
36. Vlasenko GG (1979) Electrochemical investigations of sulphide–chloride melts. PhD (Kandidate) Thesis, Institute of General and Inorganic Chemistry, Kiev
37. Mon'ko AP, Andriiko AA (1999) Ukrainian Chem J 65:111–118
38. Andriiko AA, Mon'ko AP, Lysin VI, Tkalenko AD, Panov EV (1999) Ukrainian Chem J 65:108–114
39. Orel VP (1982) Studies on the reactions of quartz and alkali silicates with hydroxide–salt melts. PhD (Kandidate) Thesis, Institute of General and Inorganic Chemistry, Kiev
40. Dmitruk BF, Dubovoi PG, Zarubitskii OG (1983) Zhurnal Prikladnoi Khimii (USSR J Appl Chem) 56:1634–1637
41. Andriiko AA, Delimarskii YK, Tchernov RV (1979) Poroshkovaya Metellurgiya (USSR Journal "Powder Metall") 5:8–14
42. Prutskov DV, Andriiko AA, Tchernov RV, Delimarskii YK, Khvalin AP (1983) Ukrainian Chem J 49:845–852
43. Prutskov DV, Pirozhkova VP, Khvalin AP (1983) Ukrainian Chem J 49:1027–1032
44. Afanas'ev PB, Zel'dovich YB, Todes OM (1949) Zhurnal Fizicheskoi Khimii (USSR J Phys Chem) 23:156–165
45. Zel'dovich YB, Todes OM (1949) Zhurnal Fizicheskoi Khimii (USSR J Phys Chem) 23:180–191
46. Delimarskii YK, Andriiko AA, Tchernov RV (1979) Ukrainian Chem J 45:1237–1242
47. Poincaré H (1885) Acta Math 7:259–338
48. Andronov AA, Vitt AA, Khaikin SE (1959) Theory of oscillations. Fizmatgiz, Moscow (English translation: Andronov AA, Vitt AA, Khaikin SE (1966) Theory of oscillators. Pergamon, London)
49. Hopf E (1942) Ber Math-Phys Kl Sächs Akad Wiss Leipzig 94:3–22
50. Nikolis T, Prigogine I (1977) Self-organization in non-equilibrium systems. Wiley, New York, NY
51. Prigogine I (1980) From being to becoming. Freeman, New York, NY
52. Prigogine I, Stengers I (1984) Order out of chaos. Bantam Books, New York, NY
53. Murrey JD (1983) Lectures on nonlinear differential equation models in biology. Mir, Moscow
54. Whitney H (1955) Ann Math 62:374–396
55. Arnold VI (1990) Catastrophe theory. Nauka, Moscow
56. Thom R (1969) Topological models in biology. Topology 8:313–352
57. Thom R (1975) Structural stability and morphogenesis. Benjamin WA, Reading, MA
58. Arnold VI, Varchenko AN, Gusein-Zade SM (1982) Singularities of differentiable mappings. Nauka, Moscow (in Russian)
59. Klimontovich YuL (1990) Turbulent motion and chaos structure: a new approach to the theory of open systems. Nauka, Moscow (English translation: Klimontovich YL (1991) Turbulence motion and Chaos structure. Kluwer, Dordrecht)
60. Ruelle D, Takens F (1971) On the nature of turbulence. Commun Math Phys 20:167–185
61. Migulin VV (1988) Fundamentals of oscillation theory. Nauka, Moscow
62. Murrey JD (1983) Lectures on nonlinear differential equation models in biology. Mir, Moscow
63. Salnikov IE (1949) Zhurnal Fizicheskoi Khimii 23:258–276

64. Svirezhev YM (1987) Non-linear waves, dissipative structures and catastrophes in ecology. Nauka, Moscow (in Russian)
65. Strogatz SH (2000) Nonlinear dynamics and Chaos: with application in physics, biology, chemistry, and engineering (Studies in nonlinearity). Perseus Publishing, Cambridge, MA
66. Rössler OE (1976) Z Naturforsch 31a:1168–1187

Chapter 2
Many-Electron Systems at Equilibrium

Any data can be generalized in an infinite number of ways. From them, we must choose only one, namely, the simplest one.

H. Poincarè

2.1 General Formulation

According to the ideas considered in Chap. 1, when electric current passes through the electrolyte containing species of a polyvalent element in highest oxidation state N, the $N-1$ low valency intermediates (LVIs) are formed in all possible oxidation states. Assuming the system is at equilibrium conditions, its properties have been determined by thermodynamic stability of the intermediates.

Naturally, the mathematical description of the equilibrium N-electron system does not require the all intervalence reactions (1.6) to be taken into consideration; a fixed set of $N-1$ equilibria will suffice for the purpose. These equilibria should be independent; that is, no reaction could be obtained by any combinations of the other ones.

The choice of the equilibrium reactions is not unique. Theoretically, several possible equivalent ways exist. The optimal choice, in our opinion, should fulfil two requirements: first, the mathematical description has to be most simple and, second, the equilibrium constants should have clear physical meaning of a measure of stability of LVIs.[1] From this point of view, a set of equilibrium reactions

[1] No possible set of independent equilibria fulfils these conditions. For example, $N-1$ elementary disproportionation reactions of Eq. (1.8) type were used to describe the reversible polarographic wave of N-electron process [1]; however, the mathematics is too cumbersome in this case.

A.A. Andriiko et al., *Many-electron Electrochemical Processes*,
Monographs in Electrochemistry, DOI 10.1007/978-3-642-35770-1_2,

$$\left(1-\frac{i}{N}\right)\mathrm{E}(0)+\frac{i}{N}\mathrm{E}(N) \rightleftarrows \mathrm{E}(i) \tag{2.1}$$

accounting for the formation of each intermediate from initial E(0) and final E(N) oxidation states, is the most convenient[2] [2].

The equilibrium constant of any reaction (2.1) becomes now a quantitative measure of thermodynamic stability of an intermediate E(i) relatively to the standard states, initial and final ones. The activity of an LVI can then be expressed in terms of the activities of the standard states and the equilibrium constants of Eq. (2.1):

$$a_i = k_i a_0^{1-i/N} a_N^{i/N} \tag{2.2}$$

and the anions of the supporting melt can contribute significantly if the species in highest oxidation state N are strong oxidants (complex fluorides of Ni(IV) or Mn (IV) [3] and other similar species could be the examples). In the Species in oxidation state i can be associated, completely or partially, as, for example in subcompounds Cd_2Cl_2 or Mg_2Cl_2. The activity of the appropriate associate containing n_i molecules (particles) can then be determined by the equation

$$a_i^{\left(\frac{1}{n_i}\right)} = k_i a_0^{\left(\frac{N-i}{N}\right)} a_N^{\left(\frac{i}{N}\right)} \tag{2.3}$$

In general case, the reactions with the molten salt solvent should also be taken into consideration. One equilibrium or more, depending on the number of possible products of reduction (or oxidation) reactions with the solvent, should be added to the basic system (2.1).

In particular, the reactions particular case when halide melts are considered, the equilibrium properties of the N-electron system can be described comprehensively by including the equilibrium

$$\frac{1}{N}\mathrm{E}(N)+\mathrm{H1}^- \rightleftarrows \frac{1}{N}\mathrm{E}(0)+\frac{1}{2}(\mathrm{H1})_2 \tag{2.4}$$

Another case is sometimes of importance when the reactions with cations of the solvent melt should be taken into account. For example, they impact considerably in a number of processes related to electrodeposition of polyvalent metals from molten salts media. The most appropriate notation for this equilibrium is proven [2] to be

[2] For example, three equilibria should be used for the description of the system where Ti metal is at equilibrium with the melt containing Ti(IV) species: $3/4\mathrm{Ti}(0) + 1/4\mathrm{Ti(IV)} \leftrightarrows \mathrm{Ti(I)}$, $1/2\mathrm{Ti}(0) + 1/2\mathrm{Ti(IV)} \leftrightarrows \mathrm{Ti\ (II)}$, and $1/4\mathrm{Ti}(0) + 3/4\mathrm{Ti(IV)} \leftrightarrows \mathrm{Ti\ (III)}$.

$$\frac{M-j}{N}\mathrm{E}(0) + \mathrm{B}(M) \rightleftarrows \frac{M-j}{N}\mathrm{E}(N) + \frac{1}{m_j}\mathrm{B}_{m_j}(j) \tag{2.5}$$

where B is the cation of the supporting melt, M is its initial oxidation number, j is the oxidation number of a subcompound formed, and m_j is the number of atoms B in oxidation state j if associates are also formed.

The activity of sub-ions b_j can then be expressed by the equilibrium constant κ_j and the activity of the solvent b_{M} as follows:

$$b_j^{\left(\frac{1}{m_j}\right)} = \kappa_j b_{\mathrm{M}} \left(\frac{a_0}{a_N}\right)^{\frac{M-j}{N}} \tag{2.6}$$

If the sub-ions B(j) are capable of further association (forming, for example, Me_2^+, where Me is an alkali metal atom [4]), then Eq. (2.6) becomes somewhat more complicated:

$$b_j = \kappa_j b_M^{q_j} \left(\frac{a_0}{a_N}\right)^{\frac{m_j(M-j)}{N}} \tag{2.7}$$

where q_j is the total number of atoms B in the associated subcompound of $\mathrm{B}_{m_j}(j) \times \mathrm{B}_{q_j - m_j}(M)$ type (for example, $Na^+ \cdot Na^0$).

If the solvent is a mixture of alkali metal halides, as often happens, Eq. (2.7) becomes more simple since $j = 0$ and, likely, $m_j = 1$:

$$b_0 = \kappa b^q \left(\frac{a_0}{a_N}\right)^{\frac{1}{N}} \tag{2.8}$$

Thus defined, such set of independent equilibria in N-electron system allows expressing the experimental data for many practically important cases in terms of a polynomial function

$$\Phi = \sum_0^N k_i X^i \tag{2.9}$$

or its logarithmic derivative

$$\overline{n} = \frac{\mathrm{d}\ln\Phi}{\mathrm{d}\ln X} = \frac{\sum_0^N i k_i X^i}{\sum_0^N k_i X^i} \tag{2.10}$$

where the variables Φ and X are calculated from experimental data.

Once such functions are obtained, they can be used for the calculation of equilibrium constants of Eqs. (2.1), (2.4), and (2.5). As one can see, Eqs. (2.9) and (2.10) are quite similar to the functions widely used in Bjerrum's theory of stepwise complex formation in solutions (Leden's and Bjerrum's functions accordingly) [5,6]. Thus, the mathematical analysis and calculation of the constants become well-elaborated routine. Some examples of such calculations are given below.

2.2 Experimental Methods: Equilibrium Conditions

2.2.1 *Solubility of Metals in Molten Salt Media*

Two main versions of the method are commonly used:

1. Constant composition of salt phase; activity of a metal in an alloy varies.
2. Activity of the metal phase is kept constant; the composition of the salt phase varies.

Taking into consideration an overall material balance of the reactions (2.1) and (2.5), the concentration of a metal which passed into the melt (that is, the solubility of metal *E*) can be written as follows:

$$S = \sum_{0}^{N-1} \left(1 - \frac{i}{N}\right)[\mathrm{E}(i)] + \sum_{0}^{M-j} \frac{m-j}{N}[\mathrm{B}(j)] \tag{2.11}$$

where [E(*i*)] and [B(*j*)] are the molar concentrations of intermediates of element *E* and sub-cations of solvent B in the melt.

Since the solubilities of metals, and hence the concentrations of intermediates are usually low, we can accept the constancy of activity coefficients and proportionality of the concentrations [E(*i*)] and [B(*j*)] to the values of activities a_i and b_j. General formulas for these activities are given by Eqs. (2.3) and (2.6).

For the case (1), $a_N = \text{const}$ and $b = \text{const}$. Thus, we can write for the solubility *S*:

$$S = \sum_{0}^{N-1} \left(1 - \frac{i}{N}\right) k_i a_0^{n_i(1-i)/N} + \sum_{0}^{m-1} \kappa_j a_0^{m_j(M-j)/N} \tag{2.12}$$

or more simple, if the alkali metal halide melt is used as a solvent:

$$S = \sum_{0}^{N-1} \left(1 - \frac{i}{N}\right) k_i a_0^{n_i(1-i)/N} + \kappa a_0^{1/N} \tag{2.13}$$

where the formal equilibrium constants are accurate to a constant multiplier with regard to the true values.

Consequently, the dependence of solubility on the activity of the metal represented in terms of the dependency $S(a^{1/N})$ takes the form of a polynomial function (2.9). Mathematical treatment of this function (graphic method of successive extrapolations, numerical least square method, or others [6]) allows for calculation of oxidation numbers, relative stability, and relative content of LVI formed.

However, two complications must be pointed out.

First: Taking into account the possible formation of solvent's sub-ions and associates, the number of terms in the polynom $S(a^{1/N})$ is of the order of $N + M$ or more. Thus, the number of experimental points in a series should be essentially more than $N + M$ if some reliable conclusions are envisaged. Meanwhile, most of experimental data reported (see, for example, review [4]) are restricted to 3–5 points only.

Second: Strictly speaking, the unambiguous conclusions cannot be obtained principally if the version (1) alone is used. As follows from Eq. (2.12), the compounds of different physical nature sometimes contribute equally to the function S. For example, in the particular system Mg(alloy)—$MgCl_2$–KCl, the linear term (with regard to $a_{Mg}^{1/2}$) corresponds to MgCl as well as to K_2Cl formation; hence, these two processes are inseparable by means of the method considered. The same is also true for the formation of physical atomic solution of metal magnesium and subcompound Mg_2Cl_2, etc.

The version (2) of the solubility method implies that $a_0 = 1$ or constant. If the equilibriums with the melt are neglected and the activity coefficients of the intermediates are constant, we can obtain the following function of the activity a_N of metal salt in the salt melt:

$$S = \sum_{0}^{N-1} \left(1 - \frac{i}{N}\right) k_i a_N^{n_i i/N} \tag{2.14}$$

that is, again, the polynomial function of $a_N^{1/N}$. More complicated equation, with negative exponent terms, accounts for the reactions with the melt [Eq. (2.5)]:

$$S = \sum_{0}^{N-1} \left(1 - \frac{i}{N}\right) k_i a_N^{n_i i/N} + \sum_{0}^{N-1} \kappa_j b_M^{q_j} a_N^{m_j (j-M)/N} \tag{2.15}$$

or, for the solution in a molten mixture of alkali metal halides,

$$S = \sum_{0}^{N-1} \left(1 - \frac{i}{N}\right) k_i a_N^{n_i i/N} + \kappa b^q a_N^{-1/N} \tag{2.16}$$

Again, the equations above show that this version of the experiment also gives rise to an ambiguity in the interpretation of the results. For example, if $n_i \neq 1$ (the

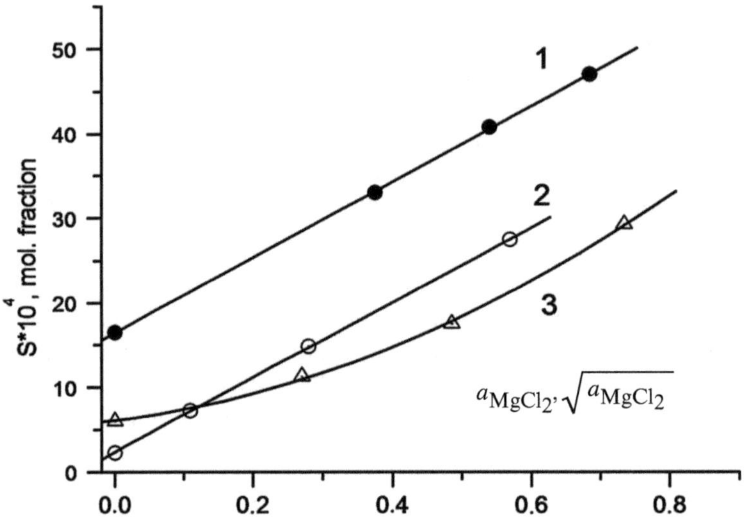

Fig. 2.1 Solubility of magnesium in molten mixtures $MgCl_2$–$CaCl_2$ (1), $MgCl_2$–KCl (2), and $MgCl_2$–NaCl (3) as a function of activity (1) and square root of the activity (2, 3) of $MgCl_2$

associates form) and $N > 2$, the formation of dimers in oxidation state 1 and monomers in oxidation state 2 cannot be separated. Hence, unambiguous conclusions can be drawn only by use of both experimental versions together. The requirements to the number of experimental points remain the same; besides, reliable data on the activities (or on the excessive thermodynamic functions [7,8]) of the components are essential, both for metal and salt phases.

No experimental data for solubility of metals in molten salts were found in the literature which would satisfy the above demands completely. As an example, we shall consider the data of Bukun and Ukshe [9] on the solubility of magnesium in the systems Mg/$MgCl_2$–MCl obtained by second version of the experiment, that is, variation of the content of $MgCl_2$ in salt melt, the activity of Mg in metal phase being kept constant [1]. The values of activities used were calculated from partial excessive free energies given by Markov [7,8] and Grjotheim et al. [10].

Some results are summarised in Fig. 2.1 and Table 2.1.

The following conclusions can be made from these data. Dimer Mg_2Cl_2 is the main product formed in $MgCl_2$–$CaCl_2$ and $MgCl_2$–LiCl systems. The monomer MgCl represents the basic state of LVI in $MgCl_2$–KCl and $MgCl_2$–KCl–NaCl systems. Both the monomer and dimer coexist in comparable quantities in $MgCl_2$–NaCl system. Also, the calculated polynomial function (2.14) includes zero-order term for each system (perhaps, except for $MgCl_2$–KCl), which can be attributed to the presence of metal magnesium in form of physical atomic solution.

It should be noted, however, that the same data could also be represented in terms of the model where sub-ions of alkali metals are formed (Eq. 2.16), with no term of zero order (see the Table 2.1). With given errors and number of the experimental data (four points per series), neither of the schemes is preferable.

Table 2.1 Analysis of the data [9] on the solubility of magnesium in molten systems $MgCl_2$–MCl at 800 °C in terms of Eqs. (2.14) and (2.16); $X = a_{MgCl_2}^{1/2}$, $b = a_{MCl}$

	Approximating equations ($S \times 10^4$ mol fraction) and calculated values of the coefficients					
	$a_0 + a_1X + a_2X^2$			$\alpha_0 bX^{-1} + \alpha_1 X + \alpha_2 X^2$		
System	a_0	a_1	a_2	α_0	α_1	α_2
$MgCl_2$–$CaCl_2$	16.5 ± 0.5	–	44.7 ± 1.2	–	–	–
$MgCl_2$–LiCl	16	5	28.4	14	0	38
$MgCl_2$–NaCl	6.1 ± 1.4	10.2 ± 8.8	29 ± 11	1.5	18	30.3
$MgCl_2$–KCl	2.3 ± 0.3	44.3 ± 0.9	–	0.4	48	0
$MgCl_2$–NaCl–KCl	13.1 ± 2.1	47.3 ± 5.8		–	–	–

Possibly, combination of the two mechanisms could be suggested; however, some additional measurements are necessary to prove this hypothesis.

2.2.2 *Potentiometry*

In the potentiometric method for study of intervalence equilibria, the potential of the metal electrode in the melt, or else, the redox potential of the inert electrode in the melt has to be measured.

Two essentially different modifications of the method are possible:

1. Measurements are taken when the electrode equilibrium is attained and no diffusion flux passes in the system.
2. Stationary values of the potential are measured, the concentrations at the electrode surface and in the bulk being not equal, and stationary diffusion fluxes of electroactive species being available in the system. Since the species at the surface are in equilibrium and the Nernst equation is still valid, we can denote such conditions as Nernstian.

Let us first consider the case (1) related to totally equilibrated systems.

The following technique was proposed [11] for processing the experimental data: The average oxidation number of low valence compounds, n, is first determined from the concentration dependency of the potential by means of the equation

$$\frac{nF}{RT} = \frac{\mathrm{d}\ln C}{\mathrm{d}\varphi} \tag{2.17}$$

After that the equilibrium constants are calculated from the values of conventional standard redox potentials.

In our opinion, this method has at least two significant drawbacks. Firstly, it does not consider the possibility of the reactions with the supporting electrolyte and formation of associates. Secondly, the existing methods for determining the standard potentials for equilibria involving LVIs (inflections of titration curves or concentration dependencies of the potential) are not reliable, especially if the equilibrium concentration of LVI is small. That is why the method above could give the correct results only in some simplest cases. A more precise approach is considered below.

Redox potential of an equilibrium system is related to the activities of the components by Nernst equations, which can be conveniently written here as follows:

$$\varphi = \varphi^0_{0/i} + \frac{RT}{iF} \ln \frac{a_i^{1/n_i}}{a_0} \tag{2.18}$$

for the reaction involving species E_i and

$$\varphi = \varphi^0_{j/M} + \frac{RT}{(M-j)F} \ln \frac{b_M^{q/m_j}}{b_j^{1/m_j}} \tag{2.19}$$

for the electrode equilibrium involving the sub-ions of solvent like $B_{m_j}(j) \times B_{q-m_j}(M)$.

By combining Eqs. (2.18) and (2.19) with the expressions for equilibrium constants (2.3) and (2.6), we obtain formulae connecting the potential values to the activities of intermediates and activity of one particular oxidation state chosen for a standard one, for example, a_0:

$$a_i^{1/n_i} = k_i a_0 P^i \tag{2.20}$$

$$b^{i/m_j} = \kappa b_M^{q/m_j} P^{j-M} \tag{2.21}$$

where

$$P = \exp\left[\frac{F}{RT}(\varphi - \varphi^0_{0/N})\right] \tag{2.22}$$

Also, the following relations remain true:

$$\varphi^0_{0/N} - \varphi^0_{i/N} = \frac{RT}{iF} \ln k_i \tag{2.23}$$

$$\varphi^0_{j/M} - \varphi^0_{0/N} = \frac{RT}{(M-j)F} \ln \kappa_i \tag{2.24}$$

Then, if the equilibrium potential of the metal electrode is obtained as a function of total molar concentration C_{E} of all species of the metal E in a fairly diluted solution in the supporting melt, this can be represented, at $a_0 = 1$, as a polynomial function of P:

$$C_E = \sum_0^N k_i P^{n_i i} \quad (2.25)$$

The parameters of each intervalence equilibrium can be determined from Eq. (2.25), both qualitative and quantitative as well. However, it should not be forgotten that the ambiguity still exists when the formation of associates is possible, because more than one different intermediate could contribute with the same value of $n_i i$.

Usually, the value of $\varphi^0_{0/N}$ [see Eq. (2.18)] which is necessary for the calculations is unknown. In this case, the arbitrary reference frame of potential can be chosen in such a way that zero value of φ is fixed approximately in the middle of the whole range of experimental potential values. Then, the calculated coefficients of the polynomial would contain the arbitrary factor $\exp(-\varphi_0 F/RT)$, where φ_0 is the value of standard potential in the chosen reference system. This factor could be calculated from the coefficient at P^N since, by definition, $k_N = 1$ [Eq. (2.2)].

A version of the experiment is possible when the metal E(0) is brought into contact with the melt containing only the highest oxidation state E(*N*) with defined known concentration, and the measurements are performed after establishing all equilibria in the system. In this case, Eq. (2.25) is no longer valid for the dependency of potential on the initial concentration of E(*N*). A true equation can be derived from the material balance of the equilibria (2.1) and (2.5):

$$NC_E^{in} + \sum_0^{M-1} (M-j)[B(j)] = \sum_0^N i[E(i)] \quad (2.26)$$

or, if the solvent is an alkali metal halide ($M = 1, j = 0$),

$$NC_E^{in} + [B(0)] = \sum_0^N i[E(i)] \quad (2.27)$$

Taking into account Eqs. (2.20) and (2.21), we obtain from Eq. (2.27):

$$NC_E^{in} = \sum_0^N ik_i P^{n_i i} - \kappa P^{-1} \quad (2.28)$$

Multiplying both sides of Eq. (2.28) by P, we obtain the dependence of $C^{in}P$ on P in a form of polynomial function of the power $N + 1$ with negative zero power term. It would mean that a notable reaction with solvent cations occurs.

Equation of (2.25) type turns out to be a good approximation for the data, taken from the book [11], on potential of beryllium electrode in LiCl–KCl–$BeCl_2$ melt at 527 K (Fig. 2.2). The plot of $C_{Be}P^{-1}$ on P is a straight line that does

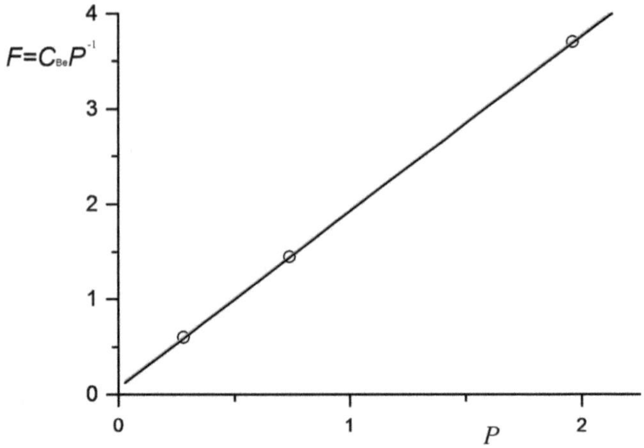

Fig. 2.2 Presentation of data [11] on the equilibrium potential of beryllium electrode in LiCl–KCl melt in terms of Eq. (2.25):

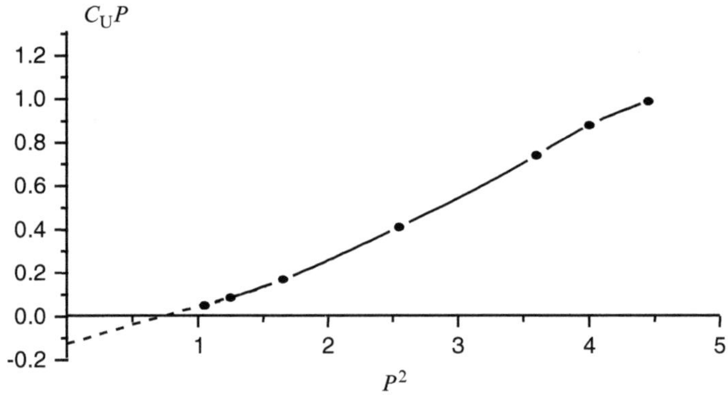

Fig. 2.3 Concentration dependence of the potential of uranium metal electrode in LiCl–KCl–UCl_3 melt [12] represented in terms of Eq. (2.28)

not pass to the origin of the coordinate system, thus evidencing the formation of a low valence chloride like BeCl. The equilibrium constant of the reaction $1/2Be + 1/2BeCl_2 \leftrightarrows BeCl$ is equal to $(4.5 \pm 0.7) \times 10^{-2}$. The data do not allow for any suggestion on the possible formation (or absence) of dimer Be_2Cl_2 because this compound and $BeCl_2$ contribute to the function (2.25) in the same way.

The potential of the uranium electrode in LiCl–KCl melt [12] can be approximated by equation of (2.28) type. The data of Fig. 2.3 clearly show the negative limit of CP dependency on P^2 when $P^2 \to 0$. This should be an evidence for essential contribution of the reaction of uranium metal with alkali metal cations. Quantitative calculations are hardly ever possible there, because the equilibrium conditions of the experiments are not quite clear.

2.3 Experimental Methods: Nernstian Stationary Conditions

2.3.1 General Notions and Validity Conditions

Eventually, the investigations of the properties of equilibrium systems aim at the determination of thermodynamic stability of LVIs by the calculation of equilibrium constants of Eqs. (2.1) and (2.5) or Eq. (2.4); after that, one can calculate the concentrations of all species in equilibrium.

Two groups of methods can be applied to this task: equilibrium, which have already been considered above, and non-equilibrium ones. The latter, in turn, can be divided into stationary and nonstationary methods. The non-equilibrium methods can be applied only to Nernstian systems, which means that the electrode equilibrium is attained. Of course, the intervalence equilibria have also to be established. The whole system, however, is non-equilibrium because of the presence of diffusion fluxes of E(i) and B(j). These fluxes are constant at stationary conditions:

$$J_i = \beta_i([\mathrm{E}(i)]_\mathrm{s} - [\mathrm{E}(i)]_{\underline{\mathrm{v}}}) \tag{2.29}$$

$$J_j = \beta_j([\mathrm{B}(j)]_\mathrm{s} - [\mathrm{B}(j)]_\mathrm{v}) \tag{2.30}$$

where $\beta_{i,j} = D_{i,j}/\delta$, $D_{i,j}$ are the diffusion coefficients, δ is the thickness of diffusion layer, and indices s and v are related to the electrode surface and bulk volume accordingly.

The current of each component, the association being allowed, is equal to

$$I_i = in_iFJ_i \tag{2.31}$$

We can express the concentrations at the electrode surface via the electrode potential using equations in form of Eqs. (2.20)–(2.22). Then the equations of stationary methods can be obtained from current balance

$$\sum_1^N I_i + \sum_1^M I_j = I \tag{2.32}$$

where the external current I, in a particular case, can be zero. If each component of the system is soluble, then the sum of fluxes must be zero:

$$\sum_0^N n_iJ_i = \sum_0^M m_jJ_j = 0 \tag{2.33}$$

Using the relationships (2.29)–(2.33) together with Eqs. (2.20) and (2.22) for concentration dependencies of the potential, one can obtain the equations representing the experimentally measured properties of an equilibrium many-electron system via the equilibrium constants for any kind of stationary methods.

However, it should not be forgotten that the final expressions would be valid only if the intervalence equilibria would have had time enough to ascertain. General physical considerations that can be used to estimate the fulfilment of this condition are as follows.

The characteristic value of a time constant for a transient process can be written in the form

$$\tau_{eq} \approx (k_e C_e)^{-1} \tag{2.34}$$

where k_e is a rate constant of the slowest elementary intervalent reaction (1.6) and C_e is the concentration of the species E in the melt. A similar time constant for attaining the steady state of diffusion is approximately equal to

$$\tau_d \approx \delta^2 D^{-1}, \tag{2.35}$$

where δ is the thickness of diffusion layer.

Hence, a method is valid when the condition

$$\tau_{eq} \ll \tau_d \tag{2.36}$$

is satisfied.

Taking $\delta \approx 3 \times 10^{-2}$ cm and $D \approx 3 \times 10^{-5}$ cm^2 s^{-1}, the value of τ_d is of order 10^2 s. The value of k_e, above the temperatures 500–600 °C, is perhaps not less than 10^4–10^5 s^{-1} cm^3 mol^{-1}. This gives the order of value for τ_{eq} in a range of 1–10^{-1} at common concentrations of dissolved species about 10^{-4} mol cm^{-3}.

For example, $k_e = 1.5 \times 10^5$ s^{-1} cm^3 mol^{-1} for the reaction 2Ge(III) $\rightarrow$ Ge(II) + Ge(IV) in KF–NaF eutectic melt at 750 °C [13], and $k_e = 2 \times 10^6$ s^{-1} cm^3 mol^{-1} for 2Si(II) $\rightarrow$ Si(III) + Si(I) in KF–KCl melt [14]. These result in the values of $\tau_{eq} = 7 \times 10^{-2}$ s and $\tau_{eq} = 5 \times 10^{-3}$ s correspondingly at $C_e = 10^{-4}$ mol cm^{-3}. It seems likely that the condition (2.36) is true for most many-electron systems in ionic melts.

2.3.2 *Stationary Potential*

Provided the concentration of E(N) in the bulk of melt is C_N, and the supporting melt is alkali metal halide(s), the equation

$$NC_N = \sum_0^N ik_i P^{in_i} - \kappa P^{-1} \tag{2.37}$$

follows from the general equations (2.29)–(2.33) describing the concentration dependencies of *stationary* potentials of a metal E electrode immersed into the electrolyte containing the highest oxidation state E(N) only. Equation (2.37) is similar to Eq. (2.28) for the dependency of equilibrium potential on the initial concentration, the numerical values of formal equilibrium constants differing by the multiplier D_i/D_N (it seems to be about 1).

If the reactions with solvent melt are neglected, the concentration dependence of the stationary potential can be represented in form of the polynomial function

$$C_N = \sum_0^N \frac{i}{N} k_i P^{in_i} \tag{2.38}$$

This formula is similar but not identical to Eq. (2.25) for the equilibrium potential.

In particular, the logarithmic derivatives of functions (2.25), n, and (2.38), $\bar{n}$, which are determined as

$$n, \bar{n} = \frac{\mathrm{d}\ln C_N}{\mathrm{d}\ln P} = \frac{RT}{F}\frac{\mathrm{d}\ln C_N}{\mathrm{d}\varphi} \tag{2.39}$$

are not equal, being related by the equation

$$\bar{n} = n + \frac{\mathrm{d}\ln n}{\mathrm{d}\ln P} \tag{2.40}$$

Since the n is a smooth ascending function, then $\bar{n} > n$ everywhere.

It is commonly accepted [11] that the derivative (2.39), in case of the potential of metal/metal ions pair, is equal to the average oxidation number of metal ions in the electrolyte. Equation (2.40) shows that, if the potentials are measured at stationary but non-equilibrium condition, the oxidation number calculated in such way would be overstated. Also, it is easy to show that if the contribution from the reactions with the supporting electrolyte is essential and Eq. (2.37) is true, the method of calculation of average oxidation number would give even larger values.

The above equations have been verified experimentally for the system silicon/chloride–fluoride melts at 660 °C [2]. The initial open circuit voltage of the cell

$$\mathrm{Si_s/KF, KCl\vdots KF, KCl, K_2SiF_6/Si_s} \tag{2.41}$$

has been measured. Hence, the experimental conditions corresponded to stationary rather than to the equilibrium state.

The experimental values of OCV are plotted in Fig. 2.4 vs. the concentrations of K_2SiF_6 in the melt. Calculating the derivative (2.39), one can obtain the average value 2.25. From the commonly accepted approach, one might come to conclusion that the species of bivalent silicon prevail in equilibrium with silicon metal. However, as follows from the more exact equation (2.37), that is not the case.

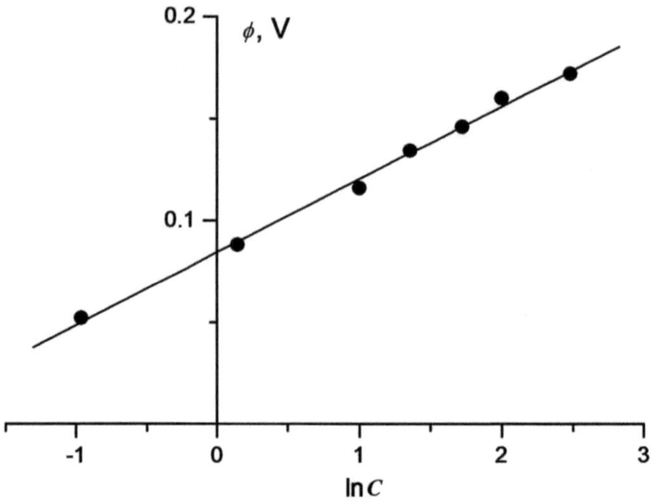

Fig. 2.4 Open circuit voltage of the cell (2.41) as a function of K_2SiF_6 concentration

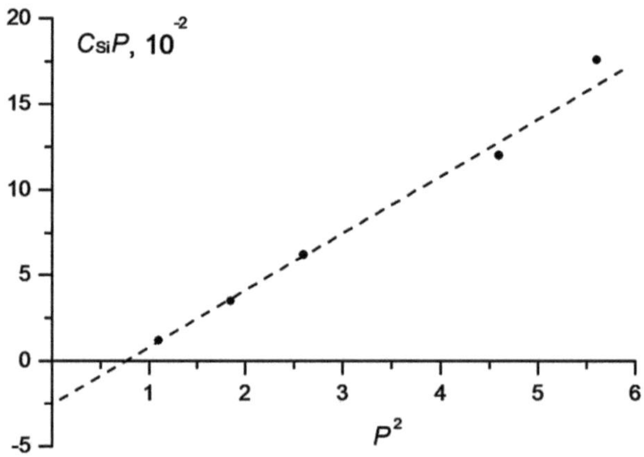

Fig. 2.5 Concentration dependence of Fig. 2.4 plotted in terms of Eq. (2.37)

Figure 2.5 represents the data as a function $C_{Si}P$ on P^2. Good linear fit is observed with the plot intersecting the negative part of ordinate axis. Hence it follows that main species in equilibrium with silicon metal are potassium sub-ions and Si(I). This conclusion consists with the data available on chemical reactions of elementary silicon and nonstationary electroreduction of silicon (IV) in chloride–fluoride melts [14].

The linear fit approximation of the data in Fig. 2.5 gives the value of equilibrium constant $\kappa = (1.34 \pm 0.84) \times 10^{-2}$ for the reaction $Si_s + K^+ \leftrightarrows Si(I) + K$ where the concentrations are molar fractions, error being calculated at confidence probability 0.95.

2.3.3 Voltammetry at Stationary Conditions (Polarography)

We shall consider now the version of stationary methods when $I \neq 0$ in current balance equation (2.32). From Eqs. (2.29) to (2.32), together with Eqs. (2.20) and (2.21), the general equation for current–potential dependency follows that is valid for diluted solutions, the alkali metal halides being used as a solvent:

$$I = \beta F \left(\sum_{1}^{N} ik_i a_0^{n_i} P^{in_i} - NC_{\mathrm{E}} - \kappa P^{-1} \right) \tag{2.42}$$

where $\beta = D_{\mathrm{N}}/\delta$; C_{E} is the concentration of element E species in the bulk in form of higher valence compound E(N), and a_0 is the activity of E(0) in zero or lowest possible oxidation state.

As follows from Eq. (2.42), a stationary current–potential curve contains the information on intervalence equilibria provided the validity condition (2.36) is satisfied. In particular, when E(0) is insoluble, then $a_0 = 1$ and the formal equilibrium constants can be obtained directly from Eq. (2.42) applying appropriate numerical processing of this function. Otherwise, if E(0) is a soluble product, then a_0 should be determined from Eq. (2.33), which in this case takes the form

$$\sum_{0}^{N} k_i a_0^{n_i} P^{in_i} = C_{\mathrm{E}} \tag{2.43}$$

Thus, Eq. (2.42) at $a_0 = 1$ describes the process without depolarization, while the system of Eqs. (2.42) and (2.43) describes the process with depolarization if the method of stationary current–potential curves is applied to the sufficiently diluted solutions of E(N) compounds.

One can see that, in the general case, the equations of voltammetric curves cannot be reduced to the simple functions of (2.9) or (2.10) type. Also it is clear that, if $\kappa \neq 0$, these dependencies are not the equations of electrochemical waves having neither anode ($P \to \infty$) nor cathode ($P \to 0$) constant limit current. It should be only noted that the curves represented by Eq. (2.42) or equation system (2.42) and (2.43), at the values of κ not too large, show the inflection point gradually transforming into the plateau of limit current while $\kappa \to 0$.

The equations for voltammetric curves become more simple when the association and reactions with the supporting electrolyte can be neglected ($n_i = 1, \kappa = 0$). For instance, a polynomial function is valid for the process without depolarization

$$\frac{I_{\mathrm{d}} - I}{I_{\mathrm{d}}} = \frac{1}{NC_{\mathrm{E}}} \sum_{0}^{N} ik_i P^i \tag{2.44}$$

where

$$I_{\mathrm{d}} = \lim_{P \to 0} I = -\beta FNC_{\mathrm{E}} \tag{2.45}$$

is the cathode limiting current.

The equation for the process with a soluble product (with depolarization) can then be presented in the form

$$\frac{I_{\mathrm{d}} - I}{I_{\mathrm{d}}} = \frac{1}{N} \frac{\sum_{0}^{N} ik_i P^i}{\sum_{0}^{N} k_i P^i} \tag{2.46}$$

Sometimes other equivalent notation of Eq. (2.46) is more convenient:

$$\frac{I}{I_{\mathrm{d}}} = \frac{1}{N} \frac{\sum_{0}^{N} jk_i P^{-j}}{\sum_{0}^{N} k_i P^{-j}} \tag{2.47}$$

where the new sum index is $j = N - i$.

Hence, the equation of $\bar{n}$-function type (2.10) describes the polarographic current–potential wave of a many-electron electrochemical process with depolarization. For example, it was established [2] that this kind of equation provides good approximation for the electrochemical wave of reduction of Si(IV) on a platinum electrode in KCl–KF–K_2SiF_6 melt (Fig. 2.6).

Following method of calculations was used. The equation of the wave was represented in form of Eq. (2.47):

$$\frac{I}{I_{\mathrm{d}}} = \frac{1}{N} \frac{\sum_{0}^{N} jk_i X^j}{\sum_{0}^{N} k_i X^j} \tag{2.48}$$

where $X = P^{-1} = \exp(-\varphi F/RT)$, φ being the value of potential in an arbitrary reference system. Then, taking into account the relation $zN = (\mathrm{d}\ln\Phi/\mathrm{d}\ln X)$, a numerical integration was carried out in order to calculate the function

$$\Phi = \sum_{0}^{N} k_i X^j = \exp\left[\frac{NF}{RT} \int_{\varphi}^{\varphi} Z\mathrm{d}\varphi\right] \tag{2.49}$$

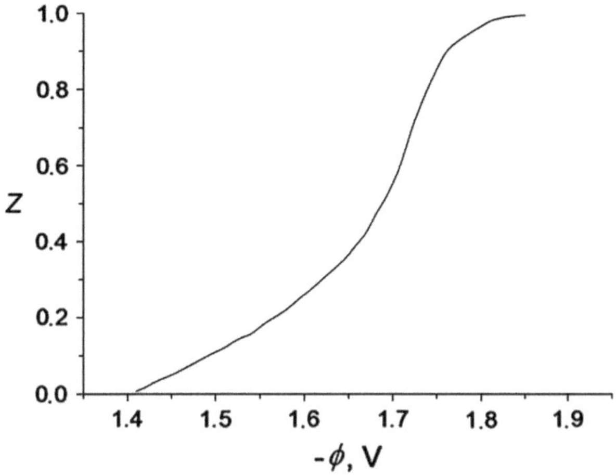

Fig. 2.6 Experimental polarogram of Si(IV) reduction on Pt electrode in KCl–KF–K_2SiF_6 at 660 °C [2]

where φ is the running value of the potential, and φ_a is the upper (anodic) limit of the integral, which, being theoretically equal to $+\infty$, is practically chosen as the potential value where the current is zero within the limits of experimental error.

Thus obtained, the values of function $\Phi(X)$ were then fitted with a polynomial of power N. Since the total number of electrons was unknown beforehand, the calculations were performed for three numbers of N equal to 2, 3, and 4. Figure 2.7 shows the best fit obtained for $N = 3$. This means that the reduction of Si(IV) to Si(I) is the overall process corresponding to the curve in Fig. 2.6. The approximating polynomial function is calculated as

$$\Phi = 1 + k_3\lambda X + k_2\lambda^2X^2 + \lambda^3X^3$$

where λ is a constant determined by the choice of zero point in the scale of potential; see the discussion on usage of Eq. (2.25); k_3 and k_2 are the formation constants of intermediates Si(III) and Si(II), the reference oxidation states being Si(IV) and Si(I).

The equilibrium constants k_3 and k_2 calculated from the polynomial fit, Fig. 2.7, are equal to 1.72 ± 0.16 and 0.23 ± 0.35, where the error limits are taken at a confidence level 0.95. From these data, one can see that the reduction of Si(IV) results in the final product Si(I). Unexpectedly, the intermediate Si(III) turned to be much more stable than Si(II).

Generally, this accords with the data available on the nonstationary process [14]. Hence, the electrochemical reduction of silicon (IV) species in halide—hexafluorosilicate melts is accompanied by formation of rather stable

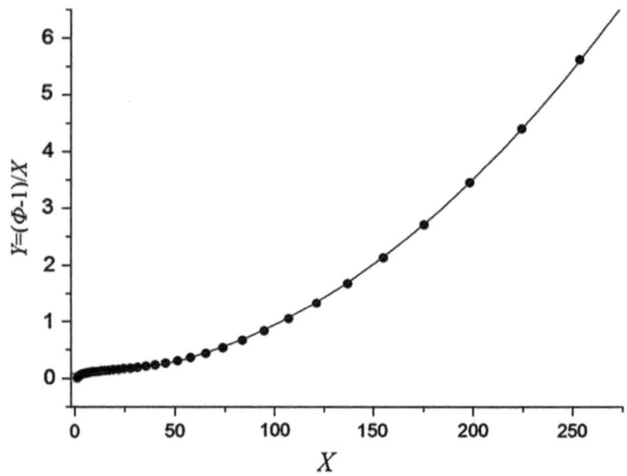

Fig. 2.7 Experimental data of Fig. 2.6 represented in form of polynomial function (2.49) (*points*), and fit by third order polynomial equation (*line*)

intermediates in oxidation states (III) and (I); the presence of Si metallic in the system onsets the reaction with alkali metal cations (see Sect. 2.3.2). In other words, the formation potential of Si metal seems to be "more negative" than the discharge potential of the supporting electrolyte. Why then the silicon can be obtained from these systems with a significant current efficiency? The concept regarding the formation of a cathode film system gives an answer. It will be discussed below in more details.

Analysis of the equations of stationary voltammetric curves (or polarographic waves) allows for a new look at the old problem of one-step many-electron waves. Following considerations should be taken in mind when evaluating the numerous experimental data on the apparent one-stage many-electron reactions:

1. If a polarographic wave of an N-electron process in terms of Heyrovský–Ilkovich equation ($\ln(I - I_d/I)$ vs. φ) has an average pre-logarithm coefficient $b > (RT/NF)$, it does not necessarily mean the electrochemically irreversible direct N-electron transfer reaction with a transfer coefficient α satisfying the relation $b = (RT/\alpha NF)$. One can see that the average value of $b > (RT/NF)$ is true for a reversible stepwise N-electron process as well.
2. Likewise, if the Kolthoff–Lingane equation $\varphi = (RT/NF)\ln(I_d - I) + \text{const}$ does not fit an experimental curve, it does not necessarily mean the depolarization due to the solubility of the final product $E(0)$ in electrolyte or electrode material, because the equation for N-electron process without depolarization has the more complex form (2.42) or (2.44).
3. Variation of shape of a many-electron wave (e.g., a half-wave potential) with concentration of depolarizer is no longer a proof of electrochemical irreversibility, since this is the case for a reversible many-electron process as well, if the formation of associated intermediate species [see equation system (2.42) and (2.43)] is taken into account.

2.4 Experimental Methods: Nonstationary Nernstian Conditions

2.4.1 Applicability

As far as the intervalent equilibria are concerned, the nonstationary methods are less informative with respect to equilibrium than the stationary methods which we have just considered. Stringent requirements to experimental conditions have to be met in order to obtain the correct data. On the one hand, these requirements are imposed by the conditions which are necessary to achieve the equilibrium for intervalence reactions, and on the other hand, by the necessity to maintain the regime of linear semi-infinite diffusion that only allows for a theoretical analysis. The combinations of the necessary conditions can be rendered by the inequalities:

$$\tau_{eq} < t < \tau_c, \tau_r \tag{2.50}$$

where τ_{eq} is determined by Eq. (2.34); t is an operation time (for example, transition time in chronopotentiometry); τ_c and τ_r denote the characteristic time values where the effect of convection (τ_c) and non-linearity (cylindricity or sphericity) of diffusion (τ_r) becomes essential.

Fixing the acceptable distortions (or errors) at a level 2 %, we can estimate the upper transition time when the impact of convection is negligible, from the relation [15]:

$$0.24(D\tau_c)^{2/3}\delta^{-3} < 0.02 \tag{2.51}$$

Taking the order of magnitude for diffusion coefficient $D \approx 10^{-5}$ cm^2 s^{-1} and for the thickness of diffusion layer $\delta \approx 3 \times 10^{-2}$ cm, we have got the estimation $\tau_c \approx 14$ s.

Sometimes, for example in highly corrosive media like fluoride melts, it is difficult to apply a plain surface working electrode. Thus, the common practice is to use an immersed metal wire, that is, cylinder shape electrode. To estimate the influence of non-linearity in such cases, the numerical values are useful that were first tabulated in Evans and Price [16] and are also given elsewhere [15]. It follows that the non-linearity (especially cylindricity), rather than convection, imposes the upper limits of operation times in such experiments. In particular, for cylinder wire of 0.5 mm diameter, the $\tau_r \approx 1.5$ s.

2.4.2 Chronoamperometry

If the relation (2.50) is maintained, then a relatively simple equation for calculation of the equilibrium constants can be derived for the method of chronoamperometry, where the current–time response is analysed after the potential step is applied.

Similarly to Eq. (2.29) for a steady diffusion flux, the equation for the flux of species E(i) can be written as follows [17]:

$$J_i = \sqrt{\frac{D}{\pi t}}\left([\mathrm{E}(i)]_\mathrm{s} - [\mathrm{E}(i)]_\mathrm{v}\right) \tag{2.52}$$

where the subscripts s and v mean, as usual, the electrode surface and the bulk volume of the electrolyte.

Equations (2.31)–(2.33) also remain valid. Thus, to the constant multiplier, the equations of voltamperic curves (2.42)–(2.46), should be used after substitution $I\sqrt{\tau}$ instead of I.

In particular, when the association of the species is negligible ($n_i = 1$), the final product is insoluble ($a_0 = 1$), and only the highest oxidation state is present in the bulk volume, the resulting equation takes the form

$$\frac{I}{F}\sqrt{\frac{\pi t}{D}} = \sum_{1}^{N} ik_iP^i - NC_N - \kappa P^{-1} \tag{2.53}$$

if the supporting electrolyte is based on alkali metal halides, and the reaction with solvent cations is taken into consideration.

If reactions with the ions of the supporting electrolyte are ignored ($\kappa = 0$), E(0) is soluble, and other conditions being the same, we come to the equation similar to Eq. (2.47) of a polarographic wave:

$$\frac{I}{FC_N}\sqrt{\frac{\pi t}{D}} = \frac{\sum jk_iP^{-j}}{\sum k_iP^{-j}} \tag{2.54}$$

The chronoamperometry method has not yet been used for the study of intervalence equilibria.

It should be noted that the theory accounting the formation of LVI is still very poorly developed. Some results have been obtained with regard to the chronopotentiometry method [18].

2.4.3 Chronopotentiometry

A two-electron process was theoretically considered in Andriiko and Tchernov [18]

$$\mathrm{X} \xrightarrow{e^-} \mathrm{Y} \xrightarrow{e^-} \mathrm{Z} \tag{2.55}$$

with the reversible intervalent reaction

$$X + Z \rightleftarrows 2Y \tag{2.56}$$

for two cases of insoluble and soluble final product Z.

The equilibrium constant of Eq. (2.56) is taken to be a measure of stability of the intermediate Y. It is equal to

$$K = \frac{a_Y^2}{a_X a_Z} \tag{2.57}$$

for soluble Z, and

$$K = a_Y^2 / a_X \tag{2.58}$$

if $a_Z = 1$ (insoluble Z).

All reaction steps are regarded as reversible. Then the Nernst's equation for both first and second stage of the reaction (2.55) can be written as below:

$$\varphi = \varphi_{X/Y}^0 + \frac{RT}{F} \ln \frac{a_X^s}{a_Y^s} = \varphi_{Y/Z}^0 + \frac{RT}{F} \ln \frac{a_Y^s}{a_Z^s} \tag{2.59}$$

or

$$\varphi = \varphi_{X/Y}^0 + \frac{RT}{F} \ln \frac{a_X^s}{a_Y^s} = \varphi_{Y/Z}^0 + \frac{RT}{F} \ln a_Y^s \tag{2.60}$$

for the reaction of deposition of solid Z on the inert electrode. Indices s referring to the activities of species at the electrode surface will be further omitted for the sake of brevity. Also, the concentrations instead of activities are used in the equations below, which is a fair approximation for sufficiently low concentrations of the depolarizer used in the method of chronopotentiometry.

General equation has been derived in Testa and Reinmuth [19], which relates the values of surface concentrations of different depolarizers to the time variable in a process of simultaneous discharge of several species. For our purpose, it takes a form

$$2C_X^0 D_X^{1/2} - \frac{2it^{1/2}}{\pi^{1/2} F} 2C_X D_X^{1/2} + C_Y D_Y^{1/2} \tag{2.61}$$

where t is the time from the beginning of the electrolysis and i is the current density applied.

If we denote

$$\sigma = \left(\frac{t}{\tau}\right)^{1/2} \tag{2.62}$$

where

$$\tau^{1/2} = C_X^0 F \pi^{1/2} D_X^{1/2} i^{-1} \tag{2.63}$$

and assume the diffusion coefficients of all species to be equal, then Eq. (2.61) becomes more compact:

$$2C_X^0(1-\sigma) = 2C_X + C_Y \tag{2.64}$$

From Eqs. (2.57)–(2.60) and (2.64), the theoretical relationships can be obtained for the time–potential dependencies.

Let us first consider the process without depolarization ($a_Z = 1$), for example, the deposition of metal Z on an inert electrode surface.

Just after the current is switched on, the solid phase of product Z is not yet formed and the equilibrium condition (2.58) does not hold. Thus, the whole process proceeds in two stages—before and after the product Z appears on the electrode surface. The first stage is described by simple equation of one-electron wave with the transition time being equal to 0.25τ from Eq. (2.63):

$$\psi = \frac{F}{RT}(\varphi - \varphi_{X/Y}^0) = \ln\frac{1-2\sigma}{2\sigma} \tag{2.65}$$

Following the appearance of Z, the equilibrium distribution of concentrations is established according to the condition (2.58). Solving Eqs. (2.58), (2.60), and (2.64) together, one can obtain the formula

$$\psi = \ln\frac{1}{4}\left[\sqrt{1+4b(1-\sigma)} - 1\right] \tag{2.66}$$

where

$$b = 4C_X^0/K \tag{2.67}$$

and K is the equilibrium constant (2.57).

The value of the parameter b defines the intersection point of Eqs. (2.66) and (2.67) according to the equation:

$$\sigma_c = (\sqrt{1+b} - 1)/b \tag{2.68}$$

Thus, the total chronopotentiometric curve should be described by the system of equations

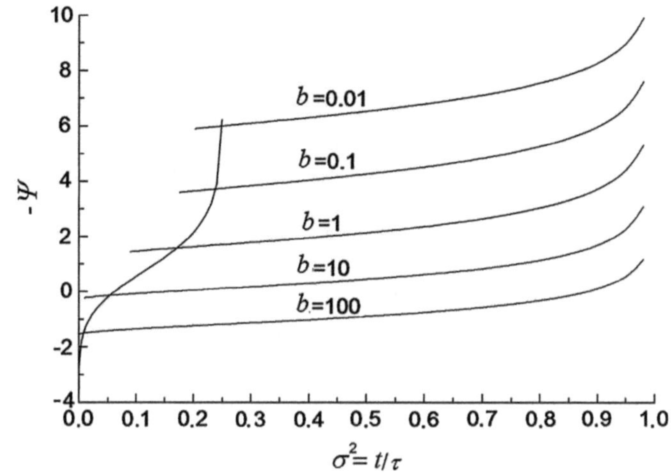

Fig. 2.8 Theoretical potential–time curves for two-electron process (2.55) and (2.56) with insoluble final product at different values of the parameter b, Eq. (2.69). Sharp maxima can appear at the intersection point σ_c because of the nucleation energy required for the formation of new phase

$$\begin{cases} \psi_1 = \ln\frac{1-2\sigma}{2\sigma}, & \sigma \le \sigma_c \\ \psi_2 = \ln\frac{1}{4}\left[\sqrt{1+4b(1-\sigma)}-1\right], & \sigma \ge \sigma_c \end{cases} \tag{2.69}$$

Figure 2.8 shows the examples of chronopotentiograms calculated from Eq. (2.69).

Several interesting conclusions follow from the brief analysis of the equations above.

Evidently, *the process looks as a sequence of two different electrochemical steps* if the curve Ψ_2 meets the concave part of the curve Ψ_1. The condition for this is

$$b = 4C_X^0/K < 3 \tag{2.70}$$

The unity term under the sign of square root may be neglected if $b \gg 1$. Then

$$\psi_2 \approx \ln\frac{1}{2}\sqrt{b(1-\sigma)} = \text{const} + \frac{1}{2}\ln(1-\sigma) \tag{2.71}$$

which corresponds to a well-known equation for a *two-electron wave* for the process $X + 2e^- = Z$ without depolarization.

Other limiting case, $b \ll 1$, gives the equation of a *one-electron wave* for the second electrochemical step of the reaction (2.55):

$$\psi_2 \approx \ln\frac{b}{2}(1-\sigma) = \text{const} + \ln(1-\sigma) \tag{2.72}$$

Hence, the general equations (2.69) describe an undistorted two-electron wave at low stability of the intermediate ($b \gg 1$, practically, $b \geq 10^3$) and separate into a sequence of two equations of one-electron waves at high stability of the intermediate ($b \ll 1$, practically, $b < 10^{-2}$). In the first case, the product Z first appears in the time when $\sigma_c = 1/\sqrt{b}$, that is, apparently at the start of electrolysis. At high stability of the intermediate, $\sigma_c \approx 0.5$, that is, the final product begins to deposit at $t = \tau/4$. Between these extremes, in the intermediate region $10^{-2} < b < 10^3$, a distorted wave, like the intermediate one in Fig. 2.8, represents the two-electron process without depolarization.

The theory of a reversible two-electron process with diffusion transport of the product into the bulk electrolyte or electrode was first elaborated in Andriiko and Tchernov [18]. Hereafter, we shall follow the analysis given in this paper.

The derivation is based on the equation [20]:

$$D_X^{1/2} C_X + D_Y^{1/2} C_Y + D_Z^{1/2} C_Z = D_X^{1/2} C_X^0 \tag{2.73}$$

and Eqs. (2.57), (2.59), and (2.64); all diffusion coefficients are taken to be equal for simplicity. The resulting equation takes the form

$$\psi = \frac{F}{RT}(\varphi - \varphi^0_{X/Y}) = \ln\left[\frac{1 + \sqrt{1 - 4\kappa\sigma(1-\sigma)}}{4\sigma} - \frac{1}{2}\right] \tag{2.74}$$

where

$$\kappa = (K - 4)/K \tag{2.75}$$

and K is determined by Eq. (2.57).

Figure 2.9 shows the potential–time curves calculated from Eq. (2.74) at different values of the equilibrium constant K.

Let us look at the behaviour of the general equation (2.74) on variation of the value of equilibrium constant (2.57), and thus on changing the thermodynamic stability of the intermediate Y.

Unstable intermediate, $K \ll 4$. Then $\kappa < 0$, $|4\kappa\sigma(1-\sigma)| \gg 1$, the unity under the square root can be neglected, and the *two-electron wave equation* follows after the simple rearrangement:

$$\psi = \frac{1}{2}\ln K + \frac{1}{2}\ln\frac{1-\sigma}{\sigma} \tag{2.76}$$

Gradually increasing the stability of the intermediate, we obtain a distortion of the two-electron waves, as if the number of electrons would become <2.[3] Remarkably, at $K = 4$, the general equation (2.74) transforms exactly into the equation of an undistorted one-electron wave:

[3] It is worthwhile to remember the "non-system" electrochemistry operating with "αn_a-equations" which was mentioned in Chap. 1. Remaining in the framework this misconception, such situation would be dealt with regarding the process as one-step two-electron irreversible one and introducing "transfer coefficient" into the denominator of pre-logarithm coefficient of Eq. (2.76).

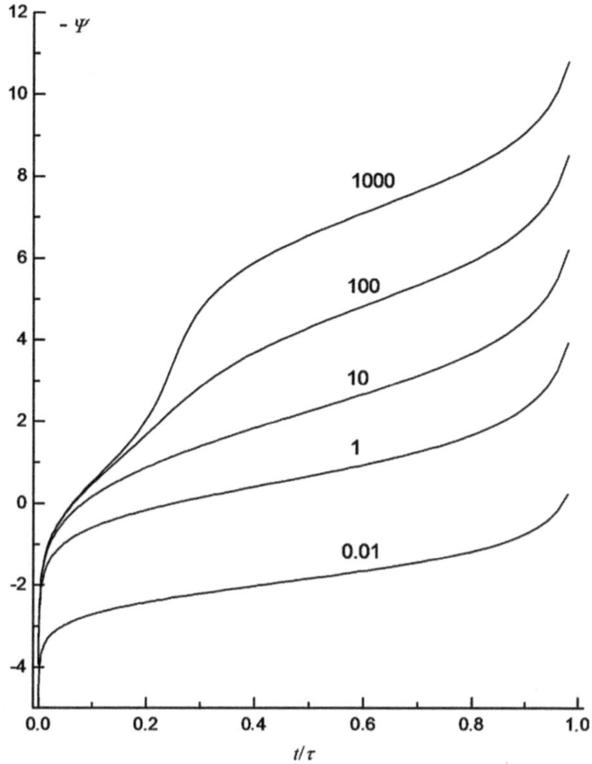

Fig. 2.9 Potential–time curves calculated from Eq. (2.74) at different values of equilibrium constant (2.57)

$$\psi = -\ln 2 + \ln\frac{1-\sigma}{\sigma} \tag{2.77}$$

Stable intermediate. Further increase of stability to $K > 4$ gives rise to the inflection point at $\sigma = 0.5$ (i.e. at $t = \tau/4$). This inflection becomes noticeable at $K > 10^2$ and visualises as a distinct transition time of the first one-electron step at $K > 10^3$.

General qualitative result of the analysis when $K > 4$ is as follows:

The initial part of the whole two-electron wave (the larger the higher is the K value) corresponds to the first step $X \xrightarrow{e^-} Y$, and the final part represents the second step $Y \xrightarrow{e^-} Z$, while the middle part is described by the general equation (2.74). The wider is the distorted part, the lower is the stability of the intermediate.

The chronopotentiometric potential–time curves are often plotted and analysed in terms of semi-logarithmic functions $\ln(\tau^{1/2} - t^{1/2})/t^{1/2} = f(\varphi)$, or, in our dimensionless notion, $\ln(1-\sigma)/\sigma = f(\Psi)$. Figure 2.10 shows a series of such dependencies calculated from Eq. (2.74).

As follows from these data, the semi-logarithmic dependency at $K \leq 10^{-2}$ is a straight line with a "two-electron" slope; thus, no evidences of intermediate

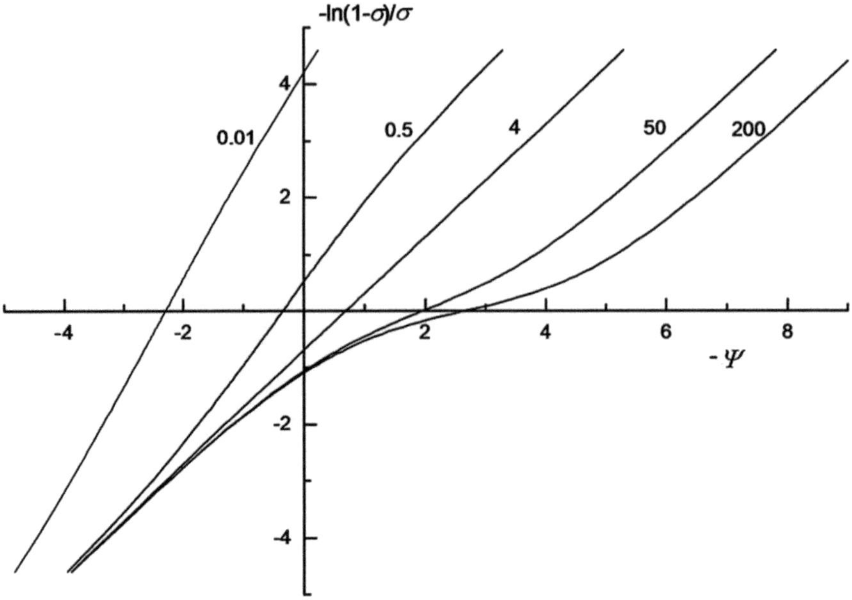

Fig. 2.10 Chronopotentiometric curves in semi-logarithmic coordinates at different stabilities of the intermediate. *Numbers* near the *curves* are the formation constant values

formation can be obtained in these conditions by the method of chronopotentiometry.[4] The dependencies become non-linear as the value of formation constant grows, with a formal number of electrons being between 2 and 1, until the plot again transforms into the straight line with "one-electron" slope at $K = 4$. After $K > 4$, the dependencies take typical S-shaped form with formal number of electrons from the slope of middle part being <1 (Fig. 2.10).

In this case, the equilibrium formation constant of the intermediate can be determined by plotting the data in "improved" semi-logarithmic coordinates ψ – ln $(1-\sigma')/\sigma'$, where $\sigma' = 2\sigma = (2t^{1/2}/\tau^{1/2})$ for $t < 0.25\tau$ and $\sigma' = 2\sigma - 1$ for $t > 0.25\tau$. It can be shown that the initial $(\sigma \rightarrow 0)$ and final $(\sigma \rightarrow 1)$ parts of the total curve may then be approximated by simple formulas $\psi = \ln(1-\sigma'/\sigma')$ and $\psi = -\ln K + \ln(1-\sigma')/\sigma'$ accordingly. Hence, the plot should contain linear initial and final "one-electron" parts, which, being extrapolated to potential axis, intercept with this axis at the distance $\Delta\varphi = (RT/F)\ln K$. Figure 2.11 gives an example of such plot, which was calculated theoretically.

Hence the shape of a chronopotentiometric wave of a Nernstian electrochemical process [Eq. (2.55)] with the reversible reaction of the intermediate formation

[4] Of course, it is correct for Nernstian reversible process, which is the only subject of this chapter.

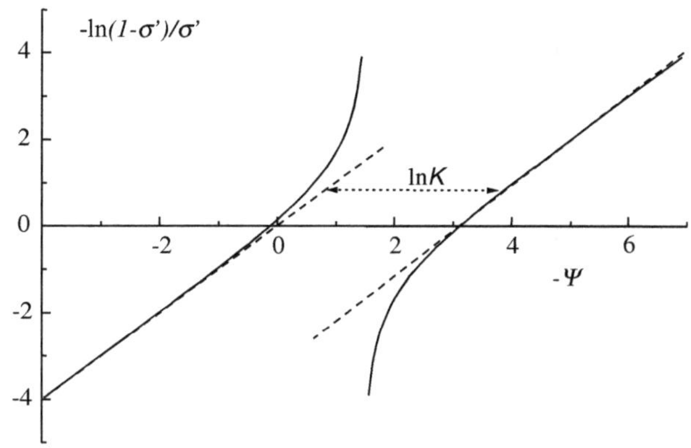

Fig. 2.11 A chronopotentiogram in "improved" semi-logarithmic coordinates at $K = 20$

[Eq. (2.56)] is determined solely by thermodynamic stability of intermediate species, which, in turn, is expressed by the equilibrium formation constant (or, alternatively, the disproportionation equilibrium constant) of the intermediate valence compound.

For the two-electron process considered, the wave should have two distinct transition times if $K > 10^3$ (for a soluble product) or $K/4C_x^0 > 10^2$ (for an insoluble product). The curves should have characteristic distortions at $10^{-2} < K < 10^3$ or $10^{-3} < K/4C_x^0 < 10^2$ accordingly. Only at lower constant values the process should look as the undistorted single two-electron step. That means, at sufficiently low thermodynamic stability of the intermediate and no kinetic limitation of the reaction (2.56), the stepwise character of the process cannot be revealed experimentally.

Thus, as we have already found out, the data of chronopotentiometric experiments cannot be presented in simple form such as Eq. (2.9) or (2.10) to calculate the equilibrium stability constants of LVIs by a standard routine. In our opinion, this method is more appropriate for the studies of intervalence reaction kinetics (see the next chapter) rather than equilibrium. However, some information on the intervalence equilibria also can be obtained in certain cases. For example, equilibrium disproportionation constant may be estimated from the shape of the wave, as was pointed out above, if $K > 4$ and the plot in semi-logarithmic coordinates is typically S-shaped as in Fig. 2.10.

Figure 2.12 gives an example of such calculations according to Delimarskii et al. [13]. It represents an experimental curve of the reduction of Ge(IV) species in the NaF–KF–K_2GeF_6 melt. The first transition time in this curve was established to correspond to the type of process like Eqs. (2.55) and (2.56), where the species X, Y, and Z are related to Ge(IV), Ge(III), and Ge(II). The values of $\Delta\varphi$ were determined at several current densities as shown in Fig. 2.12, and extrapolated to zero current. Finally, the value of $K \cong 20$ was obtained for the equilibrium constant of the reaction $\mathrm{Ge(IV)} + \mathrm{Ge(II)} \leftrightarrows 2\mathrm{Ge(III)}$ at 750 °C.

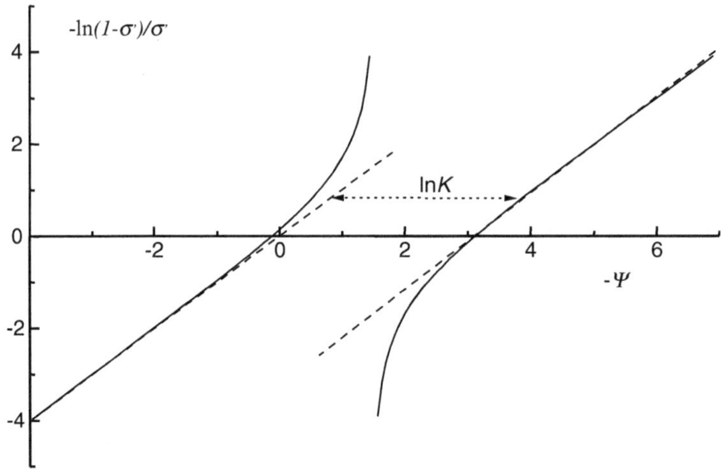

Fig. 2.12 Cathodic chronopotentiometric wave of reduction of Ge(IV) in molten NaF–KF–K_2GeF_6 plotted in "improved" semi-logarithmic coordinates. Concentration of Ge: 0.75 × 10^{-4} mol cm^{-3}, current density 3 × 10^{-2} A cm^{-2}, temperature 750 °C [13]

2.5 Instead of Conclusions: More About "System" and "Non-system" Electrochemistry

Finally, we must outline the main qualitative result of this chapter. In terms of the "system" approach outlined in Chap. 1, the following diagram can render this.

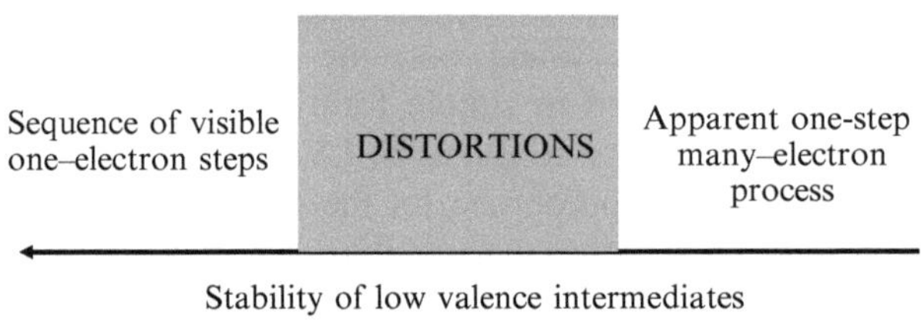

The diagram tells us that no traces of LVI in a particular experiment do not evidence against the fundamental "one-electron" approach. A lower threshold of thermodynamic stability exists where LVI cannot be registered experimentally. The magnitude of this detection threshold depends on the method applied.

There is another characteristic value of the stability where the apparent process splits into a series of sequential one-electron steps. Between these thresholds, a rather wide range of stability lies where a process manifests itself as an electrochemical wave with specific distortions.

Facing this situation, there are two ways to deal with.

The first one is "non-system" approach. The process is considered as irreversible one-step transfer of N electrons and the experimental curve is approximated by a formal electrochemical equation with a fit parameter $\alpha_N \cdot N$, the $0 < \alpha_N < 1$ being treated as the transfer coefficient of the N-electron elementary electrochemical step.

The other alternative is the "system" approach which we have adopted. Here, the correct problem consists in the determination of LVI's stability and their influence on the apparently visible process, instead of the discussion whether or not they are formed in a particular process.

What way is then really *true*? In the spirit of the philosophical ideas of Poincarè, such question is not correct. The second approach is *more useful* allowing for systematic explanation and prediction of much wider range of phenomena. In other words, the "system" approach is much more *efficient*, because it gives the uniform theoretical basis for wider range of phenomena which remain mysterious in the framework of "non-system" approach.[5]

We shall see how it works in the next chapters. Concluding this one, we have a few words to our Reader who was able to follow us through the conglomeration of the algebraic constructions.

So far we have been stayed in a quiet harbour named "equilibrium" armed with old reliable tools of classical thermodynamics. Everything was steady so far. The conformity principle was unshaken, that is, the newly derived equations comprised the former well-known relations as specific cases. Hereafter, we are going outwards this peaceful place to "non-equilibrium" ocean, where these old tools will be of less and less use.

In fact, we have already stepped on the deck—have you noticed the gentle rocking when going from totally equilibrium (solubility of metal, equilibrium potentiometry) to partly irreversible conditions (steady-state voltammetry, chronopotentiometry)? The real troubles will begin later. So, keep firmly, we take the anchor up.

[5] In a way, both approaches are true, because nobody can propose an *experimentum crusis* to distinguish which of them is right and which is wrong—such experiment is *impossible in principle*. Let us explain this statement. For example, experimental study of a process $E(\text{IV}) + 4e^- \rightarrow E(0)$ by cyclic voltammetry gave following results: (1) only one wave was observed; (2) $E(0)$ was reliably identified as the final product. OK, says a "non-system" electrochemist, the process is one-step four-electron one. We try to prove that he is wrong. For that, we carry out a second experiment investigating the same process by some modern fast technique. We apply very short pulses of current or potential and fix the intermediates by an appropriate spectral method *in situ*. Finally, we resolve each intermediate identifying it as short-lived compound which decomposes quickly by disproportionation reactions. "You see—we could say to the colleague—you are wrong, the process is stepwise, each step is one-electron, in accordance with our" system "concept, although both thermodynamic and kinetic stability of intermediates are very low and that is why we did not see it in the first experiment." "You proved nothing"—might be the answer. "The process was the one-step in the first experiment and became multistep in the conditions of the second experiment, that is all". But *the nature of process cannot be dependent on the experimental conditions!*—we oppose. Stop! Here we come to the core. The statement in italics is not a fact that can, by any means, be verified experimentally. It is pure *convention*. A fruitful convention, because it brings many separated facts into one system. That is why our concept is neither true nor wrong—it is just more convenient for any research purpose.

References

1. Boiko OI, Chernov RV (1983) Ukrainian Chem J 49:936–941
2. Andriiko AA, Boiko OI (1983) Ukrainian Chem J 53:1165–1172, 1279–1286
3. Dove MF, Court TL (1971) J Chem Soc D 14:726–735
4. Morachevskii AG, Demidov AI (1975) Reactions of metals with molten salts. In Rastvory, Rasplavy (Solutions and melts), vol 2. VINITI, Moscow, pp 242–270
5. Beck M (1973) Equilibrium chemistry of complex formation reactions. Mir, Moscow
6. Rossotti F, Rossotti C (1961) Calculation of stability constants and other equilibrium constants in solutions. McGraw-Hill, New York, NY
7. Markov BF (1974) Thermodynamics of molten salt mixtures. Naukova Dumka, Kiev (in Russian)
8. Markov BF (ed) (1985) Thermodynamic properties of molten salt systems. Reference guide. Naukova Dumka, Kiev
9. Bukun NG, Ukshe EA (1961) Zhurnal Neorganicheskoi Khimii (Soviet J Inorg Chem) 6:913–924
10. Grjotheim K, Holm JL, Malmo J (1970) Acta Chem Scand 24:77–84
11. Smirnov MV (1973) Electrode potentials in molten chlorides. Nauka, Moscow (in Russian)
12. Kisza A, Trzebiatovski W (1962) Bull Acad Polon Sci Ser Sci Chem 10:387–394
13. Delimarskii YK, Andriiko AA, Tchernov RV (1981) Ukrainian Chem J 47:787–794
14. Boiko OI, Delimarskii YK, Chernov RV (1985) Ukrainian Chemical J 51:385–389
15. Zakharov MS, Bakanov VI, Pnev VV (1978) Chronopotentiometry. Chemistry. Khimiya, Moscow
16. Evans DH, Price JE (1963) J Electroanal Chem 5:77–86
17. Damaskin BB, Petriy OA (1983) Introduction to electrochemical kinetics. Vysshaja Shkola, Moskow (in Russian)
18. Andriiko AA, Tchernov RV (1980) Ukrainian Chem J 46:1257–1264
19. Testa AC, Reinmuth WH (1961) Anal Chem 33:1320–1328
20. Baraboshkin AN, Saltykova NA, Kosikhin LT (1966) Trudy Instituta Elektrokhimii UF AN SSSR (Trans Ural Inst Electrochem USSR Acad Sci) 9:53–59

Chapter 3
Phenomenology of Electrochemical Kinetics

When studying the history of science, we observe two phenomena, which may be regarded as mutually opposite: either simplicity is hidden under the apparent complexity, or, vice versa, a visible simplicity contains extraordinary complexity in itself.

H. Poincarè

3.1 Diagnostics of Intervalent Reactions

A general way of kinetic description of a many-electron process, based on the simplified reaction mechanism (1.9), was pointed out in Chap. 1. Again, we must admit that the situation still remains complicated and a general theory of application of the experimental methods to the studies of many-electron kinetics is still poorly developed.

However, some progress has been achieved, especially with regard to the method of chronopotentiometry. The general criterion for diagnosis of the type of associated chemical reactions is among the main results in this field [1–3]. The adequate choice of the reaction mechanism is facilitated significantly by applying this criterion. It can be explained as follows.

The information on the chemical reactions associated or conjugated with electron transfer is hidden in the term W_i of the reactive diffusion equations like (1.5). This term is a function of component concentrations depending on the particular mechanisms of chemical stages and their rates.

The values of transition times, rather than the shapes of potential–time curves, contain essential kinetic information.

The well-known Sand's equation represents a particular solution of differential equation (1.5) at $W_i = 0$, that is, for pure diffusion process. It relates the transition

A.A. Andriiko et al., *Many-electron Electrochemical Processes*,
Monographs in Electrochemistry, DOI 10.1007/978-3-642-35770-1_3,

time τ and current density i to the concentration Y_0 of depolarizer in the bulk of electrolyte:

$$i\tau^{1/2} = \frac{1}{2}(\pi D)^{1/2} nFY_0 = AY_0 \tag{3.1}$$

Thus, the product $i\tau^{1/2}$ at constant values of Y_0 is a constant independent on current density.

At the onset of conjugated chemical reactions, when $W_i \neq 0$, the product $i\tau^{1/2}$ becomes a function of the applied current density. The explicit functions as a result of the exact solution of Eq. (1.5) are available only for a few simplest cases when W_i is a linear function of the depolarizer concentration, that is, only for accompanying chemical reactions of the first order.

An attempt has been made to elucidate the general effect of conjugated reactions on the shape of $i\tau^{1/2}(i)$ dependence [2]. Following considerations were used:[1]

1. If the conjugated reactions occur, then the parameter Y_0 in Eq. (3.1) would represent some average effective concentration of depolarizer dependent of the current value:

$$i\tau^{1/2} = AY(i) \tag{3.2}$$

2. The impact of chemical processes could be negligible at current values large enough; that is, $Y(i) \to Y_0$ when $i \to \infty$, Y_0 being the bulk concentration of electroactive form of depolarizer before the current is applied.
3. Evidently, $Y(i) = Y_0 + \Delta Y(i)$ and hence $Y(i) > Y_0$, if the conjugated reactions replenish the electroactive form of depolarizer in course of the process.
4. It is also clear that $Y(i) = Y_0 - \Delta Y(i)$, or $Y(i) < Y_0$, when partial inactivation of the depolarizer occurs due to the conjugated reactions.

These considerations are represented schematically in Fig. 3.1.

Although semi-intuitive, the above considerations can be summarized by the following general rule:

Providing the diffusion is planar and semi-infinite, and double layer charging effect is negligible:

- *The derivative* $\mathrm{d}(i\tau^{1/2})/\mathrm{d}i > 0$ *if the conjugated reactions of depolarizer's inactivation (auto-inhibiting) take place or the diffusion process is partially overlapped with the discharge process of surface layer (slow adsorption).*
- *The derivative* $\mathrm{d}(i\tau^{1/2})/\mathrm{d}i < 0$ *if the depolarizer is partially replenished by antecedent or subsequent (auto-catalytic) chemical reactions.*

[1] Similar, though incomplete, ideas can be found in the more recent book [4].

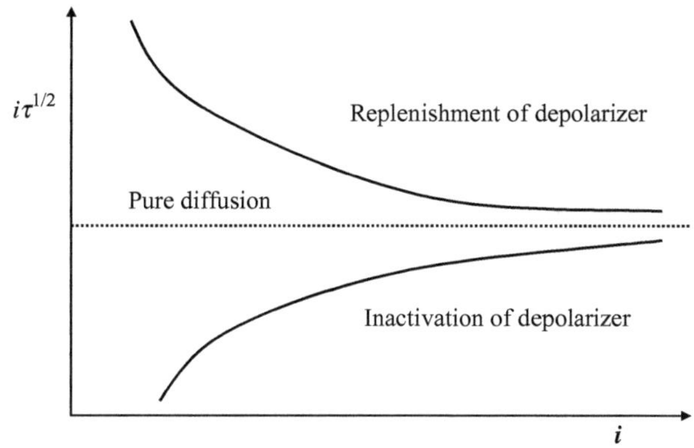

Fig. 3.1 Graphical presentation of diagnostic criterion for the type of conjugated chemical reaction for the method of chronopotentiometry

Of course, the heuristic considerations represented by Fig. 3.1 cannot prove the rule rigorously. To obtain such proof, we can formulate the mathematical problem to be solved as follows:

There is a differential equation in partial derivatives:

$$\frac{\partial Y(x,t)}{\partial t} = D\frac{\partial^2 Y(x,t)}{\partial x^2} + W(Y) \tag{3.3}$$

with the initial condition

$$t = 0; \quad x \geq 0; \quad Y = Y_0 \tag{3.4}$$

and boundary conditions

$$t \geq 0; \quad x = \infty; \quad Y = Y_0 \tag{3.5}$$

and

$$t \geq 0; \quad x = 0; \quad D\frac{\partial Y}{\partial x} = I = \text{const.} \tag{3.6}$$

Following assertion has to be established:

Taken the value of variable $t = \tau$ *so that* $Y(0, \tau) = 0$.
The derivative $\mathrm{d}(I\tau^{1/2})/\mathrm{d}I > 0$ *if, at any* x, t, *the function* $W(Y)$ *is negative.*
The derivative $\mathrm{d}(I\tau^{1/2})/\mathrm{d}I < 0$ *if the* $W(Y)$ *is positive.*

Since, physically, $W(Y)$ means a rate of chemical reaction, the problem can be discussed a little simpler—the function is an algebraic dependence on Y and, very probably, can be represented in terms of a two-parameter expression:

$$W(Y) = \pm kY^p \tag{3.7}$$

where k is a constant dependent of rate constant values of chemical reactions, and p is an apparent reaction order being commonly $0 \leq p \leq 2$.

Upon our knowledge, this mathematical problem was neither solved nor even stated before. It belongs to the mathematical physic's theory of parabolic differential equations [5], and significant mathematical difficulties are anticipated to obtain its solution. Thus, the enunciated rule on the impact of conjugated chemical reactions on the sign of the slope of $i\tau^{1/2}(i)$ plot is not strictly proved yet. Nevertheless, the available theoretical results on particular reaction schemes confirm its validity. The Table 3.1 represents the literature data on the slopes of $i\tau^{1/2}(i)$ plots for different reaction mechanisms.

Thus, the sign of the derivative $\mathrm{d}(I\tau^{1/2})/\mathrm{d}I$ can be used as a diagnosis criterion for the identification of chemical reactions conjugated with electrode process. In form of the rule above, this conclusion has first been formulated in Delimarskii et al. [2].

Similar rule is probably true for the method of linear sweep voltammetry, *if the function* $i_\mathrm{p}V^{-1/2}(V^{1/2})$ *is taken instead of* $i\tau^{1/2}(i)$, i_p and V being peak current and the rate of potential sweep accordingly. This suggestion is based on the similarities of the both methods, which were pointed out by Galus in his well-known book [21]. However, since the boundary problem of the theory of voltammetry is more complicated, the exact proof for this method should be even more difficult to obtain.

The diagnostic criterion above can be usefully employed at the first stage of the study to abridge the range of plausible reaction schemes. Further on, several techniques could be used for the elaboration of the exact (or at least most probable) reaction mechanism.

The examples below show how it works in particular experimental research results.

3.2 Electrochemical Reduction of Germanium Compounds in Fluoride Melts

The general mechanism of a many-electron process, which we consider in this book, was first employed persistently in a series of experimental works on the electrochemical behaviour of germanium species in the eutectic melt KF–NaF with K_2GeF_6 and GeO_2 additives [2, 3, 22]. The main kinetic results were obtained by the method of chronopotentiometry.

A cylindrical immersion electrode (Pt wire of 0.5 mm diameter) was used as a working electrode, because assembling of flat microelectrode proved to be very hard in corrosive media such as fluoride melts. The upper limit of working transition times was then restricted by the effect of non-linear (cylindrical) diffusion. As described in Sect. 2.4.1, it was calculated to be 1.5 s. As for the lower limit, it was imposed by the effect of double layer charging. According to [23], it was evaluated

Table 3.1 Reaction mechanism of total electrochemical process and qualitative behaviour of $i\tau^{1/2}(i)$ plots

$\frac{d(i\tau^{1/2})}{di}$	Type of process	Reaction scheme	$\lim i\tau^{1/2}$ $i \to \infty$	$\lim i\tau^{1/2}$ $i \to 0$	Function for data analysis	References
0	Pure diffusion	$Y \xrightarrow{e^-} R$	$A_Y Y_0$	$A_Y Y_0$	$i\tau^{1/2}(i)$, $\tau^{1/2}(i^{-1})$	[6]
+	Non-kinetic problems: Charging double layer capacity	$Y \xrightarrow{e^-} R$	–	$A_Y Y_0$	$i\tau(\tau^{1/2})$	[7, 8]
	Discharge of surface layer (slow adsorption, surface film)	$Y \to Y_S$; $Y_S \xrightarrow{e^-} R$	–	0	$i\tau(\tau^{1/2})$	[9, 10]
	Inactivation of depolarizer with subsequent chemical reaction (auto-inhibition)	$Y \xrightarrow{e^-} R$; $R + Y \to Z$	$A_Y Y_0$	0	$\frac{i\tau^{1/2}}{A_Y Y_0}(\lg Y_0 \tau)$	[11, 12]
–	Non-kinetic problems: (a) Non-planar diffusion	$Y \xrightarrow{e^-} R$	$A_Y Y_0$	∞	$i\tau^{1/2}(\tau^{1/2})$	[13]
	(b) Finite diffusion (effect of convection)					[14]
	Generation or regeneration of depolarizer with subsequent or preceding chemical reactions	$Z \leftrightarrow Y \xrightarrow{e^-} R$	$A_Y Y_0$	$A_Z Z_0$	$i\tau^{1/2}(i)$, $(i\tau^{1/2})^{-1}(1/\tau^{1/2}.)$	[15, 16]
		$Y \xrightarrow{e^-} R$; $R + Z \to Y$	$A_Y Y_0$	$A_Z Z_0$	$i\tau^{1/2}(i^2)$	[17]
		$Z \rightleftarrows X + Y$; $Y \xrightarrow{e^-} X$	$A_Y Y_0$	$A_Z Z_0$	$i\tau^{1/2}(i^{3/4}\tau^{1/4})$	[18]
		$Z \rightleftarrows X + Y$; $Y \xrightarrow{e^-} R$	"	"	"	[18]
		$Z \rightleftarrows 2Y$; $Y \xrightarrow{e^-} R$	$A_Y Y_0$	$2A_Z Z_0$	$i\tau^{1/2}(i^{1/3})$	[19]
		$Y \xrightarrow{e^-} R$; $2R \to X + Y$	$A_Y Y_0$	$2A_Y Y_0$	$i\tau^{1/2}(i^{2/3})$	[20]

as $\tau_c > 2.5i^{-1}$ ms; then the relative error due to the impact of charging could not exceed 2 %.

Hereby limited:

$$2.5 \times 10^{-3} i^{-1} < \tau < 1.5 \quad (\mathrm{s}) \tag{3.8}$$

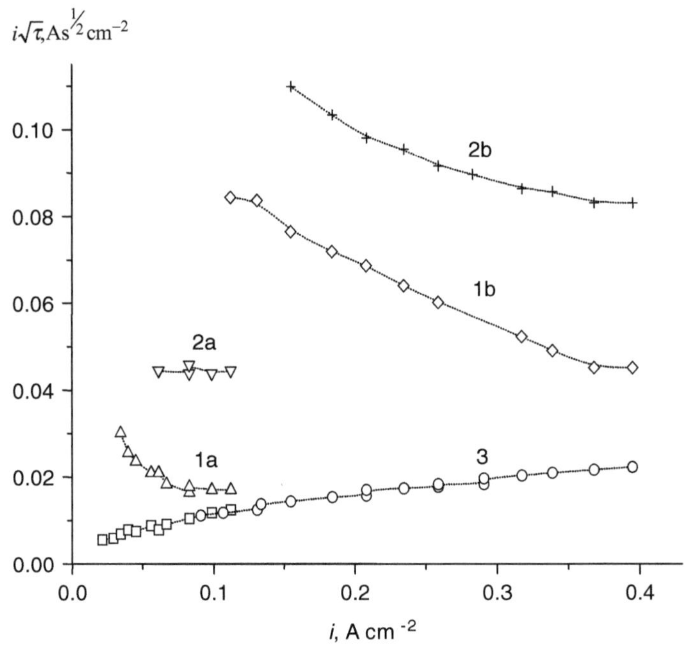

Fig. 3.2 Analysis of chronopotentiometric transition times of Ge(IV) cathodic electroreduction in KF–NaF–K_2GeF_6 melt [2]. Concentrations of Ge(IV) species: 0.752×10^{-4} (*plots a*) and 1.103×10^{-4} (*plots b*) mol cm^{-3}, temperature 750 °C

the chronopotentiometric curves expose several transition times with the values of $i\tau^{1/2}$ dependent on current density applied (see [2], for more details). Figure 3.2 shows the typical plots for an oxygen-free $(KF–NaF)_{eut}$ – K_2GeF_6 melt at 750 °C.

The transition times were determined by the position of inflection points (concave down/up). Such method was shown [24] to be most reliable in case of complicated reaction mechanisms.

Initial part of the curves includes a wave with two equal transition times of the order of several milliseconds. For this initial doubled wave, the values of $i\sqrt{\tau}$ are independent of the concentration of depolarizer and positively sloped (Fig. 3.2, plot 3). As stated above, such behaviour can be determined by the effect of imposed discharge of a surface layer.

Figure 3.3 confirms this hypothesis. The data are represented here in form of the dependence $i\tau = f(\sqrt{\tau})$. The plot shows $i\tau =$ const evidencing for a pure, diffusion-free, surface process [25]. The surface charge $Q_s = i\tau = 1.6 \times 10^{-3}$ C cm^{-2} was calculated and then the surface excess $\Gamma = Q/nF = 8.3 \times 10^{-9}$ mol cm^{-2} was determined, taken $n = 2$ because the initial process requires at least two

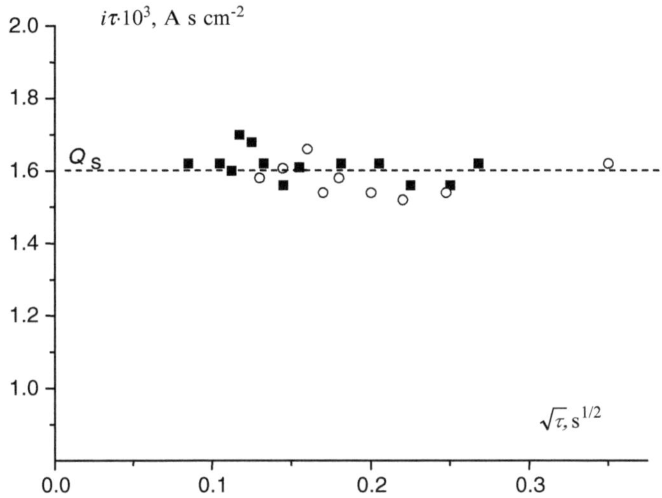

Fig. 3.3 Product $i\tau$ for initial wave 3 in the plot of Fig. 3.2

electrons. This matches well with the calculated value $\Gamma_{\rm m} = 2.5 \times 10^{-9}$ mol cm^{-2} for the adsorption of a monolayer on a platinum surface if the unevenness of the electrode is taken into account.

Since the initial waves were non-diffusion, the origin (zero time) was taken at the moment of their end while calculating the following-up transition times. Then, the rule of Testa and Reinmuth [26] was used, which says that a total transition time is independent of every intermediate step of a complex reaction mechanism and is the same as for the overall one-step process with the aggregate number of electrons if each intermediate is electrochemically active. Hence, each transition time was measured from the origin point and includes all intermediary steps.

Thus determined, the magnitude of $i\sqrt{\tau}$ (Fig. 3.2) decreases as the current increases. The diffusion-governed region exists for the plot 1 where the values are independent of current. The potential–time curves for this region give the strait lines in $E - \ln(\sqrt{\tau} - \sqrt{t})/\sqrt{t}$ coordinates with the pre-logarithm factor equal to RT/F. This means that the first electrochemical step is the reaction $\mathrm{Ge(IV)} \xrightarrow{e^-} \mathrm{Ge}$ (III). The values $i\sqrt{\tau_2}$ (plot 2) keep approximately constant with the ratio $i\sqrt{\tau_2}/i\sqrt{\tau_1} \cong 3$ at higher currents. It suggests the formation of Ge(I) as the second stable intermediate.

As it was shown above, the negative magnitude of $i\sqrt{\tau}(i)$ derivative can be caused by either previous or subsequent reactions of generation (regeneration) of the depolarizer. The previous reaction is less probable because otherwise the shape of the curves 1 and 2 in Fig. 3.2 would be the same, and this is not the case. Then the following reaction scheme ought to be accepted for the first stage:

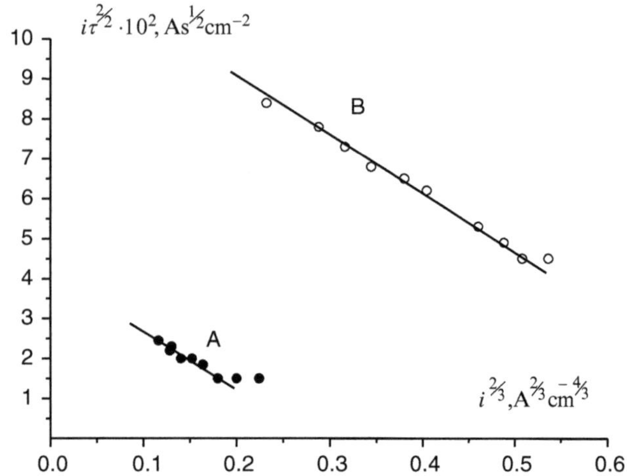

Fig. 3.4 Fischer–Dračka's [20] plots for the first chronopotentiometric waves of Ge(IV) cathodic reduction. Curves *A* and *B* correspond to the data 1*a* and 1*b* of Fig. 3.2.

$$\mathrm{Ge(IV)} \xrightarrow{e^-} \mathrm{Ge(III)}; \quad 2\mathrm{Ge(III)} \underset{k_2}{\overset{k_1}{\rightleftarrows}} \mathrm{Ge(IV)} + \mathrm{Ge(II)} \tag{3.9}$$

Fischer and Dračka [20] have considered the theory of such reaction mechanism and derived the equation for the case of a totally irreversible chemical reaction:

$$i\tau^{1/2} = C_0 F \pi^{1/2} D_{\mathrm{IV}}^{1/2} - \left(\frac{3}{16}\right)^{1/3} \pi^{1/2} D_{\mathrm{II}}^{-1/2} D_{\mathrm{III}}^{2/3} F^{1/3} k_1^{-1/3} i^{2/3} \tag{3.10}$$

where k_1 is the rate constant of Eq. (3.9) and D_{II}, D_{III} and D_{IV} are the diffusion coefficients of related species.

Figure 3.4 represents the data of Fig. 3.2 in terms of Eq. (3.10). The larger parts of these plots are linear, thus evidencing in favour of the chosen reaction scheme (3.9). The rate constant k_1 has been calculated from the slopes of these lines. For the calculations, the simplification $D_{\mathrm{IV}} \approx D_{\mathrm{III}}$ was used and the value $D_{\mathrm{IV}} \approx 0.7 \times 10^{-5}$ cm^2 s^{-1} has been determined from the level parts of the plots in Fig. 3.2. Eventually, the value of the rate constant was estimated as $k_1 = (1.5 \pm 0.2) \times 10^5$ mol^{-1} cm^3 s^{-1}. The equilibrium constant has also been calculated for this reaction (see Chap. 2, Fig. 2.12).

At the concentrations of Ge(IV) equal to or less than 0.75×10^{-4} mol cm^{-3}, the formation of Ge(II) intermediate brings about no separate transition time provided the restriction (3.8) holds true. However, the chronopotentiometric curves of the second stage of the process are typically S-shaped when being plotted as functions of the time variable $\ln\left(\sqrt{\tau_2} - \sqrt{t'}\right)/\sqrt{t'}$ where the current time is substituted by $\sqrt{t'} = \sqrt{t} - \sqrt{\tau_1}$ (see Sect. 2.4.3). This implies the formation of Ge(II) intermediate with a disproportionation constant larger than 4. The kinetics of the Ge(II) decomposition reaction manifests itself at higher concentrations (plot 2b in Fig. 3.2).

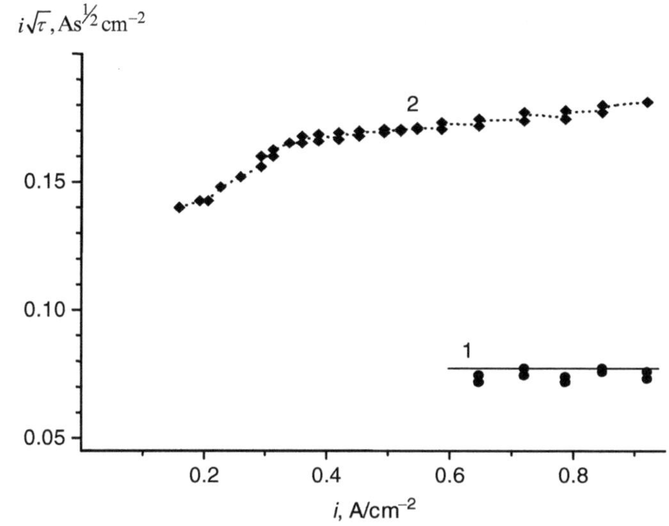

Fig. 3.5 Values $i\sqrt{\tau}(i)$ for the first (1) and second (2) waves of Ge(IV) reduction in KF–NaF–K_2GeF_6 molten electrolyte at 750 °C [3]. Total content of Ge is 4.74×10^{-4} mol cm^{-3}; the ratio $GeO_2/Ge_{total} = 0.304$

Considering the data above, the conclusion can be drawn that the general simplified scheme (1.9) approximates fairly well the mechanism of Ge(IV) discharge in the pure fluoride melt KF–NaF–K_2GeF_6. The conjugated chemical reactions other than intervalence ones do not manifest themselves in this case.

The visible pattern of the process changes when oxygen ions are added to the melt with GeO_2 (together with K_2GeF_6). If the atomic ratio of oxygen to germanium in the melt is higher than 1.5, two diffusion-governed waves are observed which can be related to the process $\mathrm{Ge(IV)} \xrightarrow{e^-} \mathrm{Ge(III)} \xrightarrow{e^-} \mathrm{Ge(II)}$. The thermodynamic stability of a Ge(III) intermediate increases as the relative content of oxygen ions increases: the equilibrium constant of the reaction $\mathrm{Ge(IV)} + \mathrm{Ge(II)} \rightleftarrows 2\mathrm{Ge(III)}$ is equal to 0.2×10^{-2}, 3.5×10^{-2} and 6×10^{-2} at O/Ge ratios 0, 1.55 and 2.0, accordingly.

If the ratio O/Ge becomes lower than 1.5, the transition time for the first step can be detected only at higher current values, and the product $i\sqrt{\tau}$ becomes current-dependent with a positive magnitude of $i\sqrt{\tau}(i)$ derivative (Fig. 3.5). According to the rule of Sect. 3.1, it could be accounted either for superposition of a surface process or for inactivation of the depolarizer. The first cause is less probable—the transition times are too large for a surface process, as, for instance, is in the case shown in Fig. 3.2 (plot 3).

The process with the inactivation of a depolarizer can be represented by the general scheme

$$\mathrm{X} \xrightarrow{\mathrm{el}} \mathrm{R}; \mathrm{X} + \mathrm{R} \rightarrow \mathrm{Z} \qquad (3.11)$$

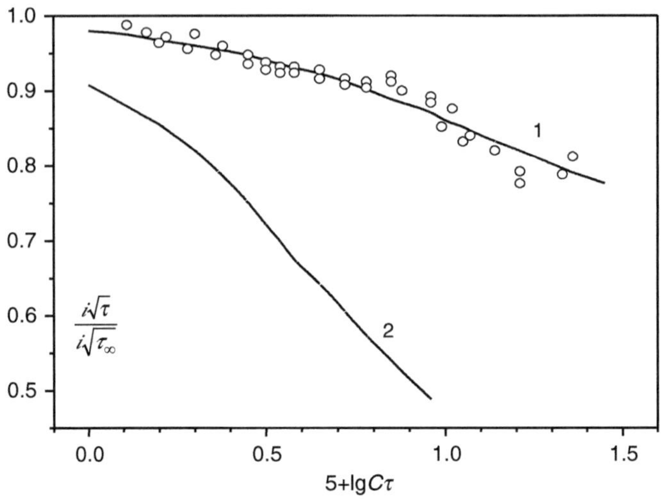

Fig. 3.6 Theoretical dependencies [12] for catalytic inactivation of the depolarizer at the electrode surface (line 1) and in the bulk of electrolyte (line 2). *Points* mark the experimental data of the plot 2 in Fig. 3.5

where el denotes the electrochemical step of the process. Of course, the mechanism (3.11) is simplified because the reverse chemical reaction is neglected and the reaction is not necessarily of the first order relative to X and R species, especially when an intermediate rather than the initial form of the depolarizer is inactivated.

The mechanism (3.11) was theoretically considered by Dracka and Fischer [12] for two cases: the surface and bulk inactivation. The problem had been solved numerically and the results were represented in terms of plots $(i\sqrt{\tau})/(i\sqrt{\tau_\infty}) = f(\lg C\tau)$ where $i\sqrt{\tau_\infty} = \lim\limits_{i\to\infty} i\sqrt{\tau}$ and C is the volume concentration of the depolarizer. The solid lines in Fig. 3.6 show these theoretical dependencies whereas points correspond to the experimental data of Fig. 3.5, curve 2. One can see that the points are scattered around the upper theoretical curve pointing to a heterogeneous reaction of inactivation. Obviously, this process corresponds to the deposition of LVI, most probably Ge(III), to form the film electrode system.

A direct experiment was carried out to check this conclusion [3, 22]. Figure 3.7 shows the stationary polarization curve of Ge(IV) reduction on a Ni electrode in KF–NaF–GeO_2–K_2GeF_6 molten electrolyte at 750 °C together with *in situ* weight change of the working electrode taken simultaneously by the method of automatic electrogravimetry developed in Delimarskii et al. [27].

In terms of Heyrovský–Il'kovich function, the part OA of the polarization curve (2), Fig. 3.7, becomes a straight line with "one-electron" slope suggesting the discharge process $\mathrm{Ge(IV)} \xrightarrow{e^-} \mathrm{Ge(III)}$. Meanwhile the electrogravimetric curve (1) comes down, which can be accounted for by "pushing up" the electrode under applied current due to electromagnetic effect as well as the change of Archimedes

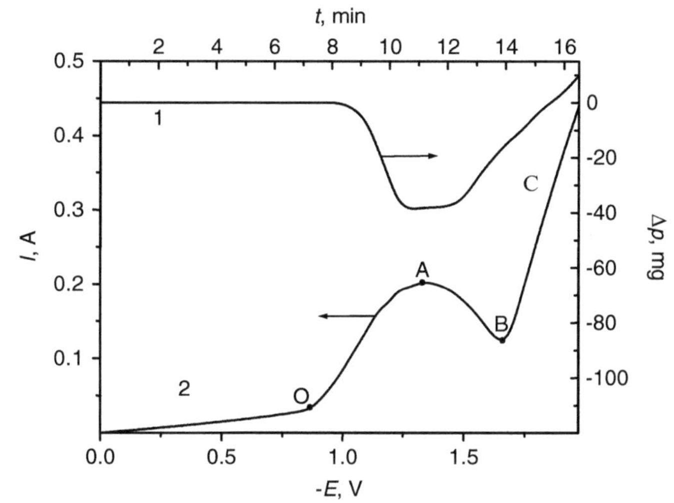

Fig. 3.7 Polarization (2) and related electrogravimetric (1) curve for the first step of Ge(IV) cathodic reduction in KF–NaF–GeO_2–K_2GeF_6 melt at 750 °C [3]. Total germanium concentration 1.3×10^{-3} mol cm^{-3}, $K_2GeF_6/Ge_{total} = 0.206$

force. The descending part AB is very similar to a polaropraphic maximum of fourth kind [28] resulting from a passivating effect of a non-metal phase deposited onto the electrode surface. The curve (1) shows that the mass of the electrode grows indeed. The Ohmic part BC points to the passivation of the electrode surface as well.

Following scheme summarizes the mechanism of the electroreduction of germanium species in molten fluorides:

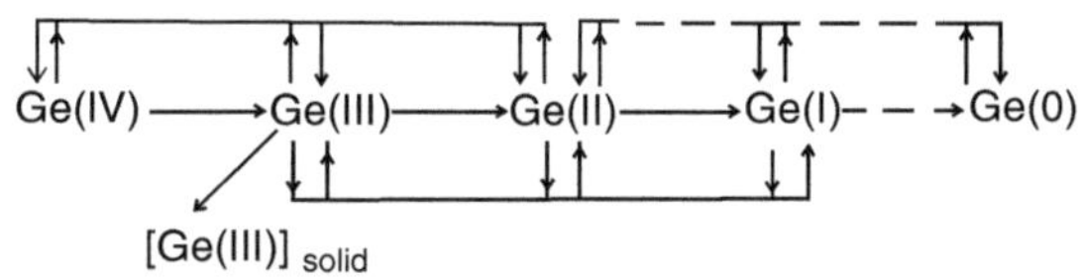

(3.12)

where $[Ge(III)]_{solid}$ is the solid LVI that deposits onto the electrode and forms the Cathode Film System. Dotted lines denote the last chemical and electrochemical stages, which occur at very negative potentials close to the discharge of the supporting electrolyte and thus may not appear in a working range of potentials.

3.3 Experimental Study of Many-Electron Kinetics: General Rules and Recommendations

The study on the electrochemical behaviour of germanium compounds in (oxy) fluoride melts had been conducted about 30 years ago [2, 3, 22]. Nevertheless, it still remains the most comprehensive example of "system" research of many-electron kinetic mechanism. Further on, we shall consider more recent examples. Before that, we should like to emphasize some general features and steps involved in such researches. They are the following:

1. Before non-stationary studies, some information is desirable to be gained about the intervalence equilibria in the many-electron system under consideration. The methods discussed in Chap. 2 can be used.
2. Most of the theoretic evaluations for complex reaction mechanisms are developed for the method of chronopotentiometry. Hence, this method should be recommended as a primary experimental technique. Of course, the other supplementary methods would be useful.
3. Correct experimental conditions and preliminary evaluations of the data are of vital importance. The effects of non-kinetic factors—charging of double layer, non-planarity and finiteness of diffusion—have to be eliminated.
4. The preliminary processing of the experimental data involves the following steps:

   (a) Preliminary determination of transition times. In case of several times inside each curve, they are measured from zero point of the curve and determined by the positions of inflecting points (concave down/up) or the relevant maxima in corresponding differential curves.

   (b) Presentation of the data in terms of $i\sqrt{\tau}(i)$ functions. If the initial wave happened to occur with a positive magnitude of the derivative, it could be attributed to some non-diffusion process, slow adsorption or discharge of surface layer. One can check this hypothesis plotting the data as a function of $i\tau(\sqrt{\tau})$. If $i\tau_s$ = const or a sloped straight line with positive intercept $i\tau_s$ on Y-axis, this proves the surface process taking place and permits to calculate its characteristics.

   (c) In case of the surface process, the subsequent transition times have to be recalculated by subtracting the initial time from the total time value. After that, the graphs $i\sqrt{\tau}$ vs. (i) are plotted again with the corrected values of the transition times.

5. After the data are preliminary handled, following routine is appropriate for further analysis:

   (a) The transition times are presented as functions of $i\sqrt{\tau}$ vs. (i), the diffusion-controlled regions where the functions keep constant are defined and the number of electrons for each of such transition time are calculated from the shapes of related potential–time curves.

   (b) The diagnostic criterion discussed in Sect. 3.2 is applied to define a class of kinetic problem and encompass all possible reaction mechanisms.

   (c) An actual reaction mechanism is elaborated with the aid of appropriate theoretical scheme available in literature. Taking this way, one can establish the type of reaction (previous or subsequent) accounting for the partial regeneration of the depolarizer if $\mathrm{d}(i\sqrt{\tau})/\mathrm{d}i, < 0$. Alternatively, if $\mathrm{d}(i\sqrt{\tau})/\mathrm{d}i > 0$, the surface or bulk inactivation of the depolarizer can be distinguished. It is useful to calculate the ratios of limit values of $i\sqrt{\tau}$ at large and small currents—they should be close to ratios of integer numbers for the accompanying intervalence reactions.

6. After the general mechanism is elaborated as stated above, the final stage has to be the determination of some quantitative parameters: rate and equilibrium constants of satellite reactions, diffusion coefficients.
7. Experimental methods other than chronopotentiometry (cyclic voltammetry, chronoamperometry, electrolysis, non-electrochemical techniques) could usefully add to the refinement of final reaction mechanism.

After the early example above, a few studies on the electrochemical behaviour of polyvalent metals in ionic melts and ionic liquids followed this line. They are reviewed in the next section and in the Chap. 6.

3.4 Some Data on the Electroreduction Mechanisms of Polyvalent Species in Ionic Melts

Cathode reduction of Si(IV) *species* in chloride–fluoride melts was studied in [29]. The general mechanism turned out to be very similar to the case of Ge(IV) electroreduction. The stepwise process was established with relatively stable Si (III) and Si(I), and less stable Si(II) intermediates. However, some specialties were found. In particular, the formation potentials of elementary silicon were unattainable being more negative than that for alkali metals of the supporting electrolyte. Together with the data on the stationary potential of the silicon electrode and stationary polarization curve (see Chap. 2), the results [29] on non-stationary kinetics bring about the following general mechanism of cathode process in KCl–KF–K_2SiF_6 melt at 700 °C:

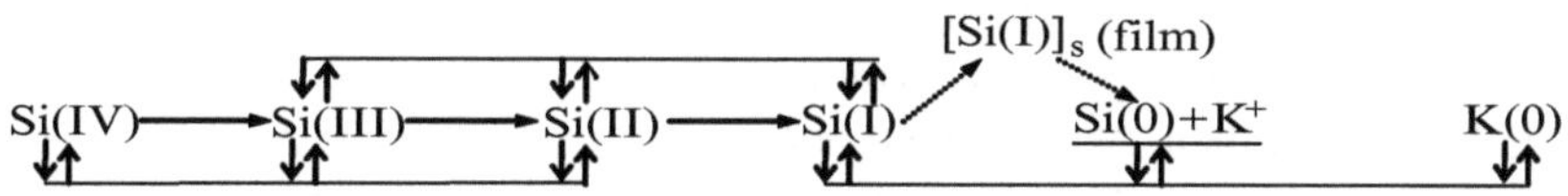

(3.13)

The species of Si(I) form at the end of the non-stationary cathode reduction. They deposit onto the electrode surface resulting in the formation of the film. Elementary silicon can be obtained only at longer times when the electrode film system with appropriate properties is finally settled down and the solid-phase reduction mechanism enters its play (see Chaps. 1 and 5). If, by any chance, the elementary silicon is available in the system, then the intervalence reaction occurs with K^+ ions taking an essential part.

Lately, the *electrochemical behaviour of uranium(IV)*, *hafnium(IV)* and *niobium (V)* was investigated in chloride melts by chronopotentiometry and linear sweep voltammetry (LSV) applying the general approach discussed above [30]. In the pure chloride melts, the formation of intermediates is even more pronounced than in fluoride-containing electrolytes, because fluoride ions stabilize the highest oxidation states and thus destabilize the intermediates.

Hence, in KCl–NaCl–UCl_4 system, the U(III) oxidation state compound turned to be the most stable intermediate. Some U(I) species were proved to deposit onto the electrode before metal formation.

Similar situation arose in the system of $HfCl_4$–NaCl–KCl. Two steps were recorded at higher concentrations of $HfCl_4$. The plot of $i\sqrt{\tau}(i)$ had a positive slope for the first step, and negative for the second step. Only at high current densities (chronopotentiometry) and at high scan rates (LSV) did the plot indicate an uncomplicated diffusion process. It was concluded that the adsorption of either Hf(II) or Hf(I) species caused the inactivation of the intermediates.

In case of the Nb(V) electroreduction in KCl–NaCl–$NbCl_5$ chloride melt at 650 °C, all one-electron steps were resolved and observed at definite conditions. The deposition of solid intermediates was identified as well.

As follows from the examples given in this chapter, general reaction mechanism (1.9), though simplified, remains very good approximation for many real processes, from high-temperature molten salts to, as we can see in Chap. 6, low temperature ionic melts (ionic liquids). We believe that, to some extent, this approach is also applicable to the processes in aqueous media.[2] However, there are obvious restrictions due to rather narrow electrochemical window of aqueous systems and solvatation effects, which would necessitate the consideration of reactions other than intervalence ones only. Sometimes, the coupled chemical reactions do not confine exclusively to the IVR even in molten salt systems. One of such examples, the electroreduction of Si(IV) species in cryolite melts, is considered below.

The impact of slow pre-dissociation was found in the process of electroreduction of Si(IV) species in Na_3AlF_6–AlF_3–SiO_2 melt at 900 °C [31, 32]. Three waves were found at SiO_2 concentrations below ~0.5 wt%, each being governed by diffusion with $i\sqrt{\tau} = \text{const}$. The first corresponds to a single two-electron reduction step $\text{Si(IV)} \xrightarrow{e^-} \text{Si(III)} \xrightarrow{e^-} \text{Si(II)}$ with an equilibrium constant of $\text{Si(IV)} + \text{Si(II)} \rightleftarrows 2\text{Si(III)} < 10^{-2}$; then, two resolved one-electron steps $\text{Si(II)} \xrightarrow{e^-} \text{Si(I)} \xrightarrow{e^-} \text{Si(0)}$ follow consecutively. At the experimental conditions applied, the kinetics of intervalence reactions had no significant impact on the apparent process, thus evidencing their high reaction rates. The measured magnitudes of Si(IV) diffusion coefficient happened to decrease as the concentration of silica in the melt increases. Physically, it could mean the presence of associated species or oligomers, which decompose as the concentration decreases. Then, one might expect the conditions where the pre-dissociation kinetics would be detectable. Indeed, this situation is observed when the SiO_2 concentrations become higher than 0.5 wt%.

---

[2] In particular, it follows from early works of Kozin summarized in [35].

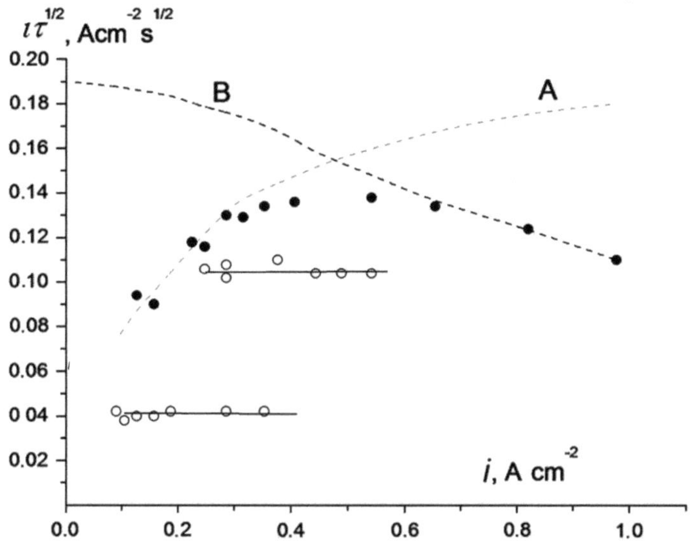

Fig. 3.8 Transition times analysis for the process of Si(IV) reduction in cryolite-based melt at 900 °C with 0.75 wt% SiO_2. *Dashed lines* show hypothetical plots for inactivation (deposition) of Si(II) (*A*) and pre-dissociation (*B*) in the scheme (3.14)

Figure 3.8 exemplifies the situation. Here, the product $i\sqrt{\tau}$ keeps constant for the first two waves attributed to stepwise discharge of Si(IV) species; however, these waves can be observed only over a narrow range of working currents. The product $i\sqrt{\tau_3}$ (last wave) behaves tricky, first rising, then attending maximum and, finally, descending. The analysis showed that the inactivation of the depolarizer due to the deposition of Si(II) species is responsible for the ascending part of the graph. The descending part is caused by slow dissociation of polymerized Si(IV) oxide species, which proceeds reversibly and shows itself only in a concentration dependence of the diffusion coefficient at lower concentrations. The scheme below summarizes the results of these studies:

$$[SiO_2]_n \rightleftarrows Si(IV) \rightarrow \{Si(III)\} \rightarrow Si(II) \rightarrow Si(I) \rightarrow Si(0), \quad Si(II) \searrow [Si(II)]_s \tag{3.14}$$

The intervalent reactions proceed reversibly and thus are not shown in the scheme. Unlike for the case of KCl–KF–K_2SiF_6 molten system, the Si(III) intermediate is least stable in the cryolite-based electrolytes. The Si(II*)* compounds are deposited onto the cathode surface to give rise to the complications for the industrial process of electrowinning Al–Si alloys [33].

The pre-dissociation reaction proceeds reversibly at low SiO_2 concentrations. The increase in silica content causes the kinetic retardation of the process. The overlapping of the both accompanying chemical reactions, pre-dissociation and inactivation of Si(II), is seen clearly at silica concentration about 0.75 wt%,

Fig. 3.8, or higher. The Si(III) intermediates are not traced in the experiment because of low thermodynamic stability in cryolite media.

A few examples above comprise almost all studies along the line of "system" approach to the electrochemical kinetics of many-electron processes in the high-temperature ionic melts. More recent data on the electrochemistry of Ti(IV) in ionic liquids are considered in Chap. 6.

Anyway, some fundamental problems are still to be overcome for wider application of these ideas. They are discussed below.

3.5 Disputable Problems of Many-Electron Kinetics

The transfer of only one electron in any elementary electrochemical step is the core idea of the developed approach. Like any other fundamental idea, by no means it is new. As it was said before, it was exploited consistently by Vetter in the 1950s of the twentieth century [34]. He considered the intermediates of successive chain of one-electron steps as surface states that neither leave the electrode to pass into the bulk of electrolyte nor undergo any chemical transformations. Based on these premises, the two-electron reaction mechanism

$$M^{z+} \underset{-e^-}{\overset{+e^-}{\rightleftarrows}} M^{(z-1)+} \underset{-e^-}{\overset{+e^-}{\rightleftarrows}} M^{(z-2)+} \tag{3.15}$$

has been considered and general equation of stationary polarization curve has been derived. These results, especially the particular cases when the rates of separate reaction stages differ essentially, become classical appearing in any electrochemical handbook.

The disadvantages of this model are evident—it cannot account for the impact of the intervalent reactions kinetics to the total electrochemical process. That this impact is significant and cannot be ignored was substantiated earlier in the works of Kozin and co-authors summarized in their book [35]. In order to allow for the contribution of the intermediates, the so-called *hetero-staged* reaction mechanism was proposed, in contrast to Vetter's mechanism defined there as the *homo-staged* one [34].

It works as follows. At least one intermediate is presupposed to appear in course of the overall z-electron process, namely M^{m+}, where $0 < m < z$. If this intermediate is formed by preceding m-electron electrochemical reaction, then the final product M^{z+} results from the subsequent disproportionation reaction:

$$M^0 \overset{\vec{k},\overleftarrow{k}}{\rightleftarrows} M^{m+} + me^-; \quad zM^{m+} \overset{\vec{k}_1,\overleftarrow{k}_1}{\rightleftarrows} mM^{z+} + (z-m)M^0 \tag{3.16}$$

Alternatively, the mechanism is allowed when the intermediate is formed by a preceding chemical reaction of reproportionation and then transforms into the final product by the electrochemical oxidation:

$$(z-m)\mathrm{M}^0 + m\mathrm{M}^{z+} \rightleftarrows z\mathrm{M}^{m+}; \quad \mathrm{M}^{m+} \rightleftarrows \mathrm{M}^{z+} + (z-m)\mathrm{e}^- \tag{3.17}$$

However incorporating the accompanying reactions with intermediate compounds, this *hetero-staged* mechanism is still uncertain, arbitrary and thus unsatisfactory and "non-system" like. Hence, the next step is obvious—it is the attempt to develop the systematic approach by synthesis of the two considered mechanisms.

The first step in this direction has already been made. The results analysed in the last chapters demonstrate the possibility of non-contradictory systematic description of many-electron systems incorporating the effect of intermediate species. However, some problems still remain to be overcome on this way. First of all, the fact of intimate coupling between electrode and intervalent reactions must be deeply understood.

This coupling, for example, follows immediately from the fact that the potential differences for separate electrochemical steps are strictly determined by intervalence equilibrium constants values; see Eq. (2.23). If the whole process is irreversible,[3] then such coupling invokes uncertainty to the determination (even definition) of the irreversible limiting step.

Let us consider, for example, a stationary non-Nernstian process. Following qualitative conclusions stems from this consideration:

1. If the IVRs occur reversibly (that is, the equilibrium relationships of 2.2 or 2.3 type are obeyed), this results immediately in the fulfilment of the Nernst equation for near-electrode space. Then, the irreversibility of an electrode reaction is not detectable in this case. In other words, the kinetics of an electrode reactions cannot show itself when the IVRs are reversible. Here lies the answer on the question why the Vetter's theory on the kinetics of stepwise electrochemical process was developed for the situation of infinitesimal rates (or absence) of coupled intervalent reactions. *No chemical kinetics—no electrochemical kinetics!*
2. If heterogeneous electrochemical steps are reversible (that is, the Nernst equation obeyed), then the conditions of chemical equilibrium for any IVR are also obeyed. Certainly, the IVRs could be irreversible as well. However, this would not reveal itself in the shape of stationary voltage–current curves bringing about only the distortions in concentration profiles of the diffusion layers near the electrode. Then, conversely to the above, *no electrochemical kinetics—no chemical kinetics* should be true as well.[4]

[3] The term "irreversibility" is used here in the narrow electrochemical sense meaning the deviation of near-surface concentrations from the equilibrium values when the Nernst' equation is no longer valid.

[4] Naturally, this reasoning is valid for stationary conditions only; the chronopotentiometric transition times, for example, depend on the rates of IVRs as discussed previously.

3. Thus, an electrochemical step could be seen as irreversible if at least one IR in the system would also proceed irreversibly. On the other hand, however, the necessary condition for electrochemical irreversibility is the irreversibility of the chemical IVR; it does not explicitly manifest itself in the stationary response of the system with irreversible electrochemical steps.

Hence, we came to some apparently paradoxical situation when the kinetics of IVR in the bulk and electrochemical electron transfer reactions at the interface happen to be inherently entangled. Then, could one disentangle a pure electrochemical step from the impact of "chemical" electron transfer reaction and study it separately? So far, this is the question with no clear answer.

References

1. Andriiko AA, Tchernov RV (1980) Ukrainian Chem J 46:1257–1263
2. Delimarskii YK, Andriiko AA, Tchernov RV (1981) Ukrainian Chem J 47:787–792
3. Andriiko AA, Tchernov RV (1981) Ukrainian Chem J 47:1202–1209
4. Kisinger PT, Heineman WR (eds) (1996) Laboratory techniques in electroanalytical chemistry, Ch 4. Marcel Dekker, New York, NY
5. Friedman A (1968) Equations in partial derivatives of parabolic type. Mir, Moscow (in Russian)
6. Sand HJS (1901) Philos Mag 1:45–57
7. De-Vries VT (1968) J Electroanal Chem 17:31–38
8. Rodgers RS, Meites L (1968) J Electroanal Chem 16:1–16
9. Herman HB, Tatwawadi SH, Bard AJ (1963) Anal Chem 35:2210–2216
10. Laitinen HA, Chambers LM (1964) Anal Chem 36:5–17
11. Shapoval VI, Reznik GV (1974) On the theory of non-stationary processes of electroreduction of anions in molten salts (chronopotentiometry). In Ionic melts, vol 2. Naukova Dumka, Kiev, pp 135–149
12. Dracka O, Fischer O (1974) Coll Czech Chem Commun 39:1970–1979
13. Mamomtov G, Delahay P (1954) J Am Chem Soc 76:5322–5332
14. Filinovskiy VY, Kirianov VA (1964) Doklady Akademii Nauk SSSR (Proc USSR Acad Sci) 156:1412–1416
15. Delahay P, Berzins T (1953) J Am Chem Soc 75:2486–2495
16. Reinmuth WH (1972) J Electroanal Chem 38:95–103
17. Delahay P, Mattax CC, Berzins T (1954) J Am Chem Soc 76:5319–5331
18. Fischer O, Dracka O, Fischerova E (1961) Coll Czech Chem Commun 26:1505–1512
19. Fischer O, Dracka O, Fischerova E (1960) Coll Czech Chem Commun 25:323–330
20. Fischer O, Dracka O (1959) Coll Czech Chem Commun 24:3046–3052
21. Galus Z (1994) Fundamentals of electrochemical analysis, 2nd edn. Wiley, New York, NY
22. Andriiko AA, Tchernov RV (1983) Electrodeposition of powdered germanium from oxyfluoride melts. In: Physical chemistry of ionic melts and solid electrolytes. Naukova Dumka, Kiev
23. Bard A (1963) Anal Chem 35:340–349
24. Baraboshkin AN, Vinogradov-Zhabrov ON (1970) Trudy Instituta Elektrokhimii UF AN SSSR (Trans Ural Inst Electrochem USSR Acad Sci) 15:126–130
25. Tatwawadi SA, Bard AJ (1964) Anal Chem 36:5–14
26. Testa AC, Reinmuth WH (1961) Anal Chem 33:1320–1328
27. Delimarskii YK, Andriiko AA, Tchernov RV (1979) Ukrainian Chem J 45:1237–1243

28. Delimarskii YK, Tumanova NK, Shilina GV, Barchuk LP (1978) Polarography of ionic melts. Naukova Dumka, Kiev (in Russian)
29. Boiko OI, Delimarskii YK, Chernov RV (1985) Ukrainian Chem J 51:385–390
30. Boiko O, Serrano K, Chamelot P, Taxil P (2000) High Temp Mater Process 4:441–450
31. Delimarskiy YK, Prutskov DV, Andriiko AA, Tchernov RV (1983) Ukrainian Chem J 49:738–743
32. Prutskov DV, Andriiko AA, Delimarskiy YK, Tchernov RV (1986) Ukrainian Chem J 51:826–833
33. Prutskov DV, Andriiko AA, Tchernov RV (1987) Tsvetnyje Metally (Non-Ferrous Metals) 2:39–42
34. Vetter K (1961) Electrochemical kinetics. Springer, Heidelberg
35. Kozin LF (1989) Electrodeposition and dissolution of polyvalent metals. Naukova Dumka, Kiev

Chapter 4
Electrode Film Systems: Experimental Evidences

What is a good experiment? This is the experiment which gives us something more than a separate fact; this is the experiment, which permits us to foresee; that is, the one which leads to generalisations.

H. Poincarè

This chapter analyses some experimental examples, demonstrating qualitatively the concept of electrode film system (EFS). The discussion is based on the classification of EFS considered in Chap. 1.

4.1 Ion–Metal Film Systems in Processes of Electrodeposition and Electrorefinement of Polyvalent Metals

The film system on metal electrodes in ionic melts often belongs to the ion–metal type. This is consistent with the general model of polyvalent conductor, which predicts the degeneration of semiconductor conductivity into metal one at higher temperatures [1, 2].

Qualitatively, the mechanism of the processes when the current passes in the cathode direction is depicted in Fig. 4.1.

The Faradaic process at the outer boundary corresponds to the partial reduction of the species with higher oxidation state to the lower valence intermediates (LVI). This process results in the deposition of a film and growth of the film with the rate being proportional to the electronic part of the total current.

The ionic component of the conductivity brings about the Faradaic process at the inner boundary of the film. It invokes further reduction of the film compounds—actually, to the phase with pure electronic conductivity. The rate of this process is proportional to the ionic current passing through the film.

A.A. Andriiko et al., *Many-electron Electrochemical Processes*,
Monographs in Electrochemistry, DOI 10.1007/978-3-642-35770-1_4,

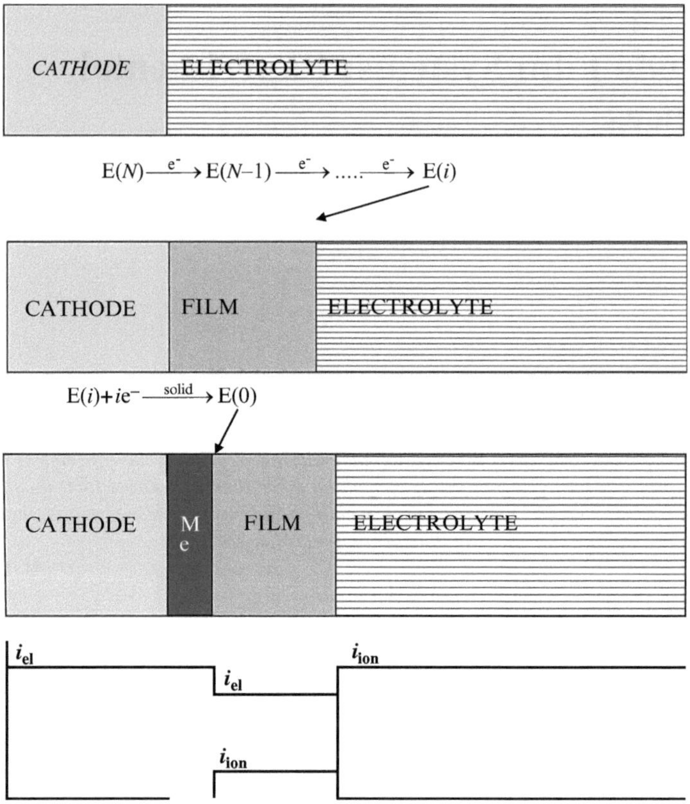

Fig. 4.1 Film system at a cathode. The distribution of ionic and electronic current components is shown schematically below

The prevailing type of charge carriers in the film should affect considerably the apparent character of the inner electrochemical process. Should these carriers be alkali metal cations, their discharge at the inner boundary would be preferable. Thus, the formation of alkali metal is expected to be the side process in such systems resulting in the low current efficiency of the targeted process of the reduction of a polyvalent metal species. Evidently, that accounts for higher current efficiency of elementary silicon electrodeposition from potassium halide baths relatively to the electrolytes based on sodium halides.

Presumably an anionic nature of ionic conductivity is favourable for the electroreduction of a polyvalent metal species in a film system. If the movable anions are liberated at the inner interface due to the Faradaic process there, this process is to be the further reduction of the compounds forming the film—actually, to the metallic phase with electronic conductivity—because the liberated anions can leave the reaction zone with the ionic current passing across the film. This is the main reason why fluoride-containing electrolytes are so widespread, especially in molten salt electrochemistry—the small fluoride anions are good carriers for the charge removing outside the surface film.

Figure 4.1 gives a formal idea about the system's behaviour in such case. There the rate of the inner electrochemical process is proportional to the ionic current. This process results in metallization of the film. The dynamics of the overall process depends, first of all, on the relative rates of the inner and outer processes. In turn, these rates are the functions of the part of the electronic conductivity of the film. When this part is large (near 1), then the compounds forming the film have to be the only final product of the electrolysis. The examples can be the electrolysis of oxide–fluoride melts containing Ti [3] or Nb [4] species, or electrodeposition of tungsten bronzes from oxide melts [5].

If the ionic conductivity is low but no longer negligible (roughly, the part of electronic conductivity is more than 0.5), the cathode deposit should consist of two layers: a thick outer one including LVIs and a thin inner one including pure metal. Such situation can be observed when some species are added to the bath increasing the electronic conductivity of the film, as it happens when copper compounds are added to the electrolyte for silicon deposition [6].

In principle, a system with several inner boundaries is possible enclosing the layers with different oxidation states and, thus, with different electronic-to-ionic conductivity ratios. This hypothetic possibility has been discussed by Vas'ko and Kovach [7].

For such a "*polysurface*" system to exist, the drop in electronic conductivity at each inner interface must be less than the drop at the neighbouring interface (closer to the electrolyte). However, the calculations showed that such situation could not be stable dynamically on sufficiently large timescales. We have so far no distinct evidence for the formation of such deposits, neither in molten nor in aqueous electrolytes.

The outlined pattern is especially important for molten fluoride systems where the transport properties of the mobile F^- ions determine the conductivity. Let us consider some qualitative details of the dynamics in such systems.

If the part of ionic conductivity is more than i/N (i is the oxidation state of LVI in the film and N is the oxidation state of initial species in the bulk of the electrolyte), the behaviour of the electrochemical film system should not be stable, as it is in the case of the non-stationary chemical systems considered in Chap. 1. In particular, it could not grow steadily with layers separated by the interface, because the rate of metallization should be higher than the rate of growth at the outer boundary. Then the stages depicted in Fig. 4.1 should alternate: after the sufficient accumulation of LVIs near the electrode, fast precipitation follows, the film thickness increases, and then the metallization of the film goes on till the electronic conductive phase "pricks" the film and enters the surface forming electronic contact to the electrode base material. After that, the discharge to the LVI takes place again, the new portion of the film precipitates, and so on.

The question arises whether these alternating stages can be synchronized in time in the whole electrode surface. Certain experimental data show that they really *can* (see below). An "oscillatory" deposition occurs then: the macroscopic parameters, like current and/or potential and the mass of the deposit, are oscillating around some averaged trend.

Other data demonstrate the dendritic (not smooth) movement of the inner boundary during the metallization of the film. Thus, the needles of dendrites pierce through the film before the metallization is completed, and then the deposit contains some amount of a non-metal phase together with the crystals of the metal product. After cooling down the deposit to the ambient temperature, the partly reduced constituent of the deposit (that is, LVI) often can be disproportionate, giving very fine metal powder dispersed between the larger primary crystals of the metal in the final product.

The above-considered pattern provides satisfactory explanation for the mechanism of formation of metal–salt cathode deposits, so-called "pears" in Russian or "carrots" in Western literature. Such deposits are very common if the electrolysis is carried out at temperatures below the melting point of a deposited metal. A number of examples can be found in the processes of electrodeposition of refractory and other polyvalent metals (Nb, Ta, Ti, Si, B and others) in fluoride-containing molten salt electrolytes.

Such electrodeposition process had been widely investigated for electroreduction of germanium species in molten $KF–NaF–K_2GeF_6–GeO_2$ electrolyte [8]. One of the most interesting information has been obtained with the automatic balance experimental technique permitting to record the weight of the electrode *in situ* in course of the electrolysis [9].

Figure 4.2 shows typical electrogravimetric curves related to the kinetics of germanium electrodeposition in galvanostatic conditions. These curves include the initial section when the weight remains almost unchanged for a few minutes. Obviously, this inductive time period is required for the accumulation of LVIs at the near-electrode area. After that the weight begins to rise with the rate higher than it is necessary for the deposition of pure germanium. This fact evidences for the co-deposition of non-metal phase as it was considered above.

The initial period of the FS formation had also been detected when the electrogravimetric and polarization curves were taken simultaneously (see Fig. 3.7). These data show the onset of deposition only after the maximum current is achieved in the polarization curve. The curve shows such maximum because of increasing ohmic resistance due to the formation of the film on the electrode; that is the reason of the decrease of the limiting current of the reduction process $Ge(IV) \overset{e^-}{\leftrightarrows} Ge(III)$.

The predicted oscillatory behaviour has been found for the potentiostatic regime of the process. Figure 4.3 shows such example when the weight of the cathode, rising in total, exhibits clearly distinct local oscillations.

Microscopic investigations of the germanium powder samples after washing out the salt components from the cathode deposit reveal two different kinds of crystals: large germanium dendrites and very fine particles. According to the above mechanism, the first can be considered as a primary electrocrystallization product, while the second is a product of a secondary disproportionation of film's LVI.

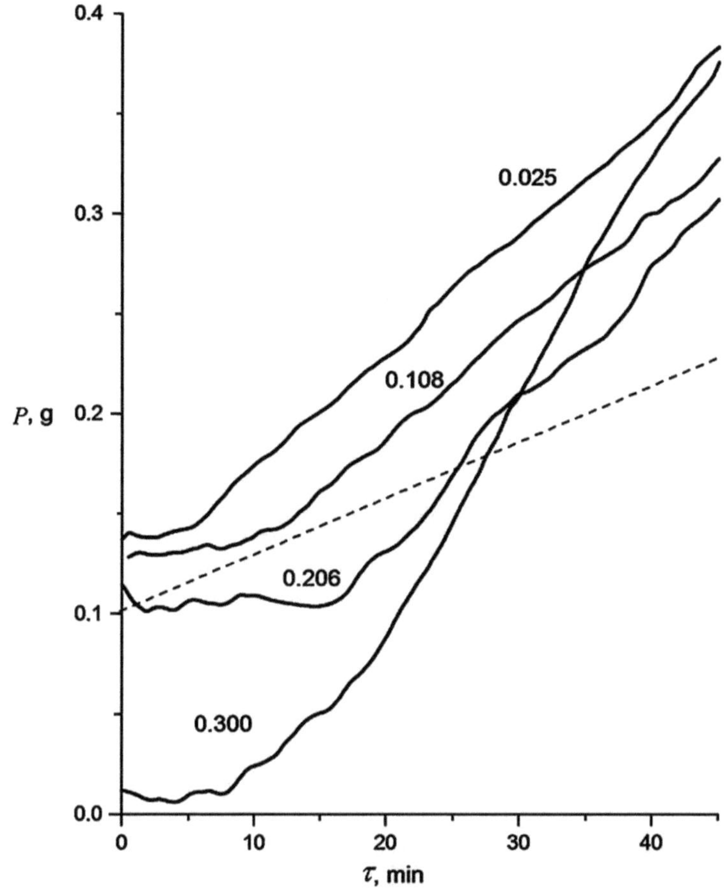

Fig. 4.2 Weight (including Archimedean force) of the cathode with the deposit vs. time in course of germanium electrodeposition from KF–NaF–K_2GeF_6–GeO_2 melts at 750 °C in galvanostatic conditions. *Dashed straight line* corresponds to the deposition of pure germanium with 100 % Faradaic efficiency [8]. The numbers indicate the ratio of GeO_2 to K_2GeF_6 in the electrolyte

Such "two-phase" composition was also found for silicon powder electrolytically produced by the method described in [10] from molten salt electrolyte KF–KCl–K_2SiF_6 with addition of silica. It is clearly seen from the plot in Fig. 4.4 representing a histogram size distribution curve obtained from the mesh analysis of a pilot-scale batch of electrolytically produced silicon.

Thus, the proposed mechanism of the processes in the ion–metal FS is in good qualitative agreement with most essential experimental facts related to the *cathodic deposition* of polyvalent metals. Let us consider now the behaviour of such system in case of a reversed (*anode*) direction of the applied current. This problem is important, in particular, for the electrolytic refinement of polyvalent metals in molten electrolytes.

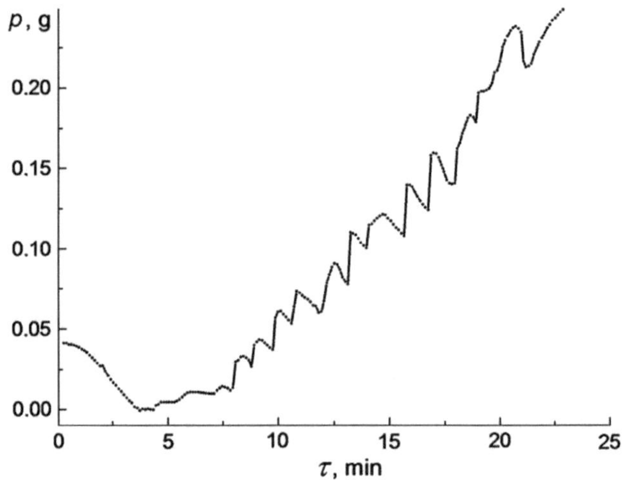

Fig. 4.3 Kinetics of potentiostatic electrodeposition of germanium from the oxyfluoride melt

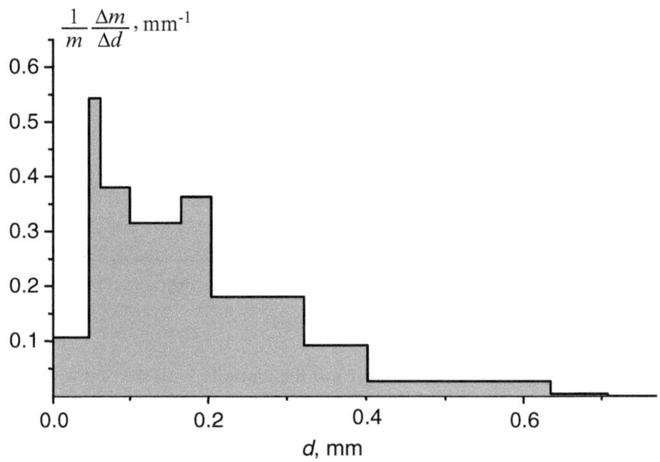

Fig. 4.4 Size distribution plot for the powder of electrolytically produced silicon, $KF–KCl–K_2SiF_6$ ($+SiO_2$) bath, 720 °C

The anodic current in the ion–metal FS brings about the oxidation of the metal to LVI at the inner boundary. These LVIs have to be oxidized at the outer boundary, passing finally to the electrolyte in form of species with higher oxidation states. This is the most general pattern for anodic dissolution of a polyvalent metal. Having this in mind, following phenomena have to be expected in course of long-time electrolysis.

If the ionic conductivity is low, the rate of film's dissolution should be higher than the rate of its formation. Thus, the film should eventually disappear and the dissolution occurs with no complications.

On the contrary, when the ionic conductivity is higher than the electronic one, the film should grow continuously. Then, after sufficiently long time of electrolysis, the complications are inevitable due to the accumulation of non-metal phases at the anode.

That is, the expected behaviour of an anodic FS with low ionic (high electronic) conductivity has to be similar to a cathodic FS with low electronic (high ionic) conductivity. This consideration leads to a rather pessimistic conclusion: practical employment of electrorefining process of a polyvalent metal in a long-time scale should appear to be a hard task if the FS would possibly form both at the cathode and at the anode.

In terms of the developed ideas, the possible complications are caused by the fact that the conditions for successful cathode and appropriate anode processes are mutually incompatible: high ionic conductivity is required for the efficient cathode deposition, whereas high electronic conductivity is necessary for easy anode dissolution.[1]

Generally, we can apprehend certain peculiarities of continuous long-term electrorefining process, which do not follow and cannot be predicted from the short-term laboratory experiments. In other words, the success of such experiments does not guarantee the success of the large-scale process.

That was true, for example, for the process of metallurgical silicon electrorefining in a KCl–KF–K_2SiF_6 melt at 750 °C. A number of short-term laboratory experiments had been carried out, excellent results were obtained and a patent had been filed [11]. Nevertheless, an attempt to reproduce these results in a large-scale process (pilot electrolyser with 500 A current load) was unsuccessful. The problems arose after operation for approximately 24 h. They manifested themselves in increasing the cell voltage, the electrolyte temperature and a drop in current efficiency. This ultimately led to technological failure and termination of the process. The cause was established to be the non-predicted slag accumulation in the anode area.

In contrast to the silicon process, durable electrorefining of Nb and Ta was more successful. The processes were performed in the molten mixture KCl–NaCl–K_2NbF_7 (K_2TaF_7) at 720 ± 10 °C in course of 1 or 2 weeks. Two different types of electrolysers were used, i.e., laboratory type and pilot plant cells, enabling operating currents up to 50 and 300 A, respectively. The techniques, other conditions and results of the experiments were described in detail in [12].

As followed from the available literature on the electrorefining of Nb and Ta, the numerous data on current efficiency, purity and other parameters differed significantly in the works of different authors. The reason was established to be very poor reproducibility of the common short-term laboratory experiment results. Some parameters of the process (e.g., current efficiency) varied significantly, especially if the electrolyte was repeatedly used in several electrolyses.

[1] The same reasoning will be valid when considering the operation of rechargeable electrode in a battery. If the FS is formed at its surface, the film must have mostly ionic conductivity during the "cathodic" part of its life (the discharge for a positive electrode and charge for a negative one), while prevailing electronic conductivity is required to behave as anode at the reversed current. Obviously, that is why numerous attempts to make a lithium metal electrode rechargeable failed.

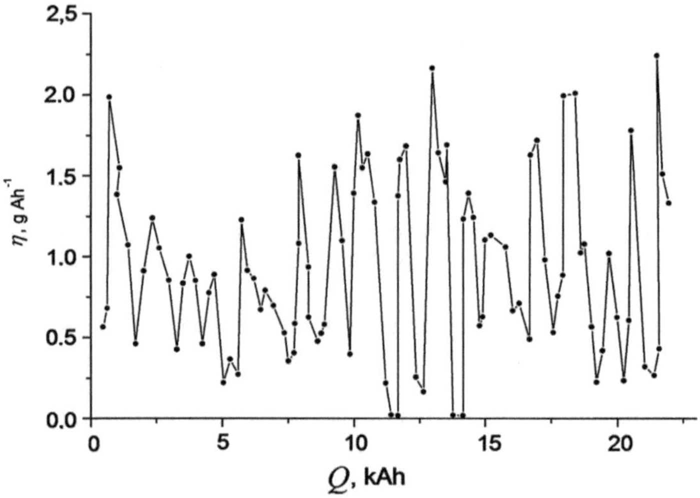

Fig. 4.5 Yield of Nb cathode deposits in a pilot-scale cell [12]. Operating current 120 A

That this irreproducibility is not a consequence of experimental errors but an effect of the inner property of the system was discovered in course of long-term continuous processes. Here, after some quantity of charge passed, each portion of the deposit was quantitatively removed from the cathode and weighed. Such obtained yields of the product per unit charge were also scattered significantly. However, this scattering was found to obey an amazing order when plotted against total time of the electrolysis (or total charge passed): the cathode current efficiency oscillated about a mean value.

Figure 4.5 shows this unusual regularity for pilot-scale process of Nb electrorefining. Similar pattern is observed for continuous electrolysis in the laboratory cell (Fig. 4.6). One can see that, though the amplitudes of the oscillations may differ significantly, the period remains constant with good precision.

Some parameters of the electrorefining of Nb and Ta in long-term electrolyse processes are presented in Table 4.1.

The theory of the observed phenomena was developed later. It is based on the above-considered ideas upon electrochemical FS and bifurcation analysis. These results will be further discussed in Chap. 5.

We are finishing now the consideration of ion–metal FS in molten electrolytes. In conclusion, one important circumstance has to be borne in mind.

In some cases, e.g., at the electrochemical reduction of silicon species, the potential of final stage (formation of a metal) turns out to be more negative than the potential of the alkali metal formation (see Chap. 3). Similar situation seems to be true for a number of other polyvalent metals. For example, the reaction of zirconium metal with fluoride melts is known [13] where the potassium metal is a product.

In such kind of systems, the winning of a pure metal is only possible by means of a solid-phase reduction mechanism, in a FS with a high ionic conductivity. Moreover, not just ionic but predominantly anionic conductivity is required. Otherwise, the cations of the alkali metals would migrate through the film to the inner interface

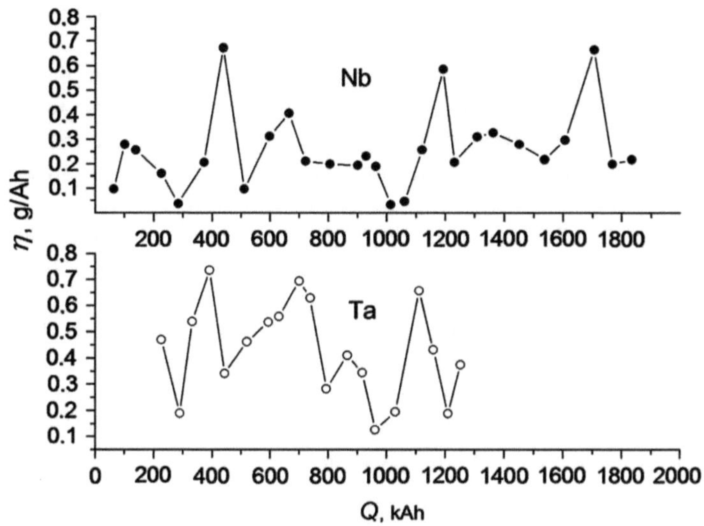

Fig. 4.6 Periodic changes of charge efficiency in long-term laboratory electrorefining of Nb and Ta at the constant current 20 A (Ta) and 29 A (Nb)

Table 4.1 Experimental parameters and results of long-term experiments on the electrorefining of Nb and Ta in long-term continuous process, electrolyte KCl–NaCl–K_2NbF_7 (K_2TaF_7), 720 ± 10 °C

	Current	Current density ($A\ cm^{-2}$)				Period T of the oscillations		TS/V[c]
Process	A	Cathode	Anode	η_{av} (%)[a]	Me (wt%)[b]	Ah	h	$10^3\ s\ cm^{-1}$
Nb, Lab	29 ± 1	0.60	0.95	32.3	36.0	258 ± 11	8.9	1.48 ± 0.11
Nb, Pilot	120 ± 10	1.20	0.60	42.3	32.0	1,262 ± 49	10.4	1.51 ± 0.18
Ta, Lab	20	0.50	0.64	26.5	40.0	239 ± 19	12.0	3.34 ± 0.25

[a]Average current efficiency calculated for the process $E(V) + 5e^- \rightarrow E(0)$ from the total charge passed for the whole run of the electrolysis
[b]Mean content of metal in the cathode deposit
[c]S is the mean geometric area of the electrodes ($\sqrt{S_{cath}S_{anode}}$), and V is the volume of the electrolyte. This number relates to some fundamental properties of the dynamic system (see Chap. 5)

discharging there to form an alkali metal. Obviously, such situation took place in the attempts to obtain elementary silicon from oxide melts—alkali metal was the main product there [14].

As it was already mentioned before, high anionic conductivity can most easily be achieved in fluoride-containing electrolytes, with mobile F^- ions as current carriers. This accounts for the widespread employment of fluoride-based electrolytes in applied electrochemistry of polyvalent metals. However, sometimes the condition of prevailing anionic conductivity is not satisfied even in fluoride-containing systems. For example, current efficiency of silicon deposition drops considerably if cations become to contribute to the ionic conductivity. It happens when sodium or, especially, lithium salts are used as bath components. The liberation of alkali metal is then the main side process.[2]

Alkali metal is a main product if the electrolysis is performed of the solutions of silica in molten fluorides with no additives of complex silicon fluorides [15]. Though the solubility of silica in such melts is high enough, it seems that the film with high anionic conductivity could form only in "acidic" melts, where the ratio O/Si is much <2.

We have seen that the idea of an *electrode film system* is useful for electrochemistry of molten salts including low-temperature ionic liquids. It is not restricted, however, to this field only. As an example, the protective layer on lithium metal in aprotic organic electrolytes could be mentioned. This layer, so-called "solid electrolyte interphase" (SEI), exhibits properties of a polyfunctional conductor with high ionic conductivity (Li ions are the carriers) and low electronic conductivity of semiconductive nature. Some peculiarities of film systems with semiconductive character of electronic conductivity are considered below.

4.2 Ion–Semiconductor Film Systems

According to Velikanov [1], the exponential growth of conductivity with temperature is the main feature of a semiconductive PFC. Taking the film system as a whole, we can recognize two other distinctions.

Firstly, the polarization characteristics should depend not only on the properties of the system, but also on the electrode material. This follows from the non-linearity of the metal–semiconductor junction, which is dependent on the work function of electrons, and thus, on the nature of the electrode material [16].

Secondly, the electrode FS should exhibit the rectifying effect caused by the potential barrier of the forbidden energy band of the semiconductor. Qualitatively, if this band is sufficiently wide, the electrons penetrating the film can reduce not only the depolarizer near the Fermi level, but also compounds of the supporting electrolyte. This is a typical situation in aqueous electrolytes where large amount of charge is usually spent for the reduction of protons or protonated species with generation of hydrogen.

It seems likely that such ion–semiconductor systems are rare in ionic melts due to the degeneracy of the semiconductor at high temperatures. However, some data evidence their formation at some peculiar conditions in course of Si metal deposition out of a KF–KCl–K_2SiF_6 melt. It was shown [17] that the initial parts of polarization curves contained the non-diffusion waves similar to that observed for germanium systems (see Chap. 3) on a platinum electrode. The same "surface" waves were detected on Ag and Au electrodes as well, thus suggesting the reduction

[2] This manifests itself in characteristic sparks and flames on the surface of the electrolyte when the process is conducted in the ambient atmosphere or in intensive evolution of hydrogen when the cooled cathode deposit is placed into water.

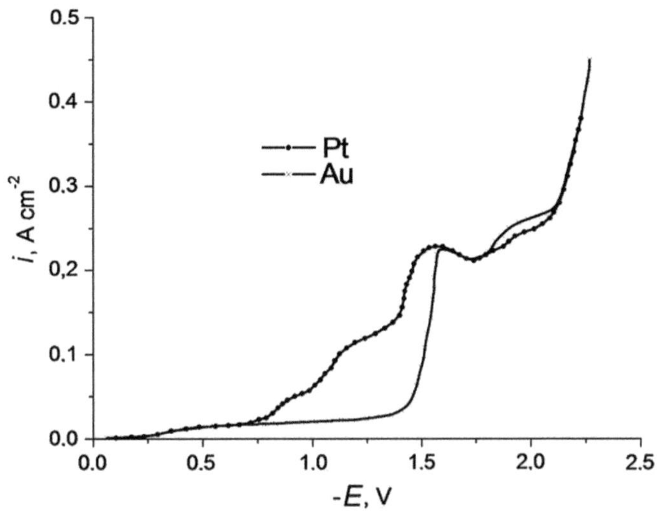

Fig. 4.7 Voltammetric curves for Si(IV) discharge in a KCl–KF–K_2SiF_6 melt on platinum and gold cathodes [17]

of chemisorbed layers of a depolarizer rather than a reaction of oxide layer of the cathode material.

Additionally, these layers (or the products of its reduction) turned to acquire the barrier effect. Their conductivity rose sharply at the potentials close to the discharge potential of alkali metal (positive by 0.2–0.4 V). This results in "cut-off" of the initial parts of discharge stepwise diffusion waves in polarization curves, e.g., at a potential −1.5 V vs. quasi-reference Pt electrode in case of Au in KCl–KF–K_2SiF_6 melt (Fig. 4.7).

In this case, the barrier properties of electrode–electrolyte interface together with the effect of electrode material can be considered as evidence for the formation of ion–semiconductor FS with low ionic conductivity.

4.3 Dielectric Film Systems

If the gap in the forbidden band of the semiconductor film is wide enough and exceeds the decomposition voltage of the supporting electrolyte, the film substance would behave as a pure ionic conductor—superionic conductor, if the ion conductivity were high.

The situation should then be favourable if the ion electric carriers would take part in the Faradaic process at the inner boundary inside the film. The ions could pass across the film discharging at the metal/film interface or, vice versa, liberating in the Faradaic process at this interface, and passing through the film into the bulk of the electrolyte. The processes at the negative electrode of a lithium battery (Li metal or Li/C intercalate compound) could be an example of such situation.

Otherwise, if the ionic carriers are electrochemically inactive or the ionic conductivity is very small, the FS would acquire qualitatively new properties becoming a dielectric, or insulating one. Some harmful phenomena, such as crisis (or blockade in terms of [18]), are indicative for such systems. They manifest themselves in sharp rise in voltage (galvanostatic regime) or drop in current (potentiostatic regime). That is, the electric resistance of a system tends to increase sharply when the dielectric electrode FS begins to develop. This brings typically the increase of temperature, spark, glow or arc discharge at the electrode, deterioration of wettability of an electrode surface, etc. Such phenomena result from blocking of a surface by a dielectric film where an appreciable current can be achieved only in conditions of a dielectric breakdown.

The so-called "*anode effect*" on carbon electrodes in fluoride-containing melts is probably the most vivid manifestation of a dielectric FS. This phenomenon was first related to the industrial electrolysis of aluminium in Hall–Heroult cells and, thus, is widely known for more than a 100 years. Nevertheless, a generally acknowledged theory is not achieved yet. Several possible causes have been considered: non-wettability of the electrode surface, electrostatic repulsion of the bubbles of the gas evolved at the anode, hydrodynamic crisis of the gas evolution and, finally, the formation of a fluorocarbon dielectric film [18–20]. This latter explanation had been developed since then, mainly by Japanese researchers, relatively to the electrochemical production of fluorine [21]. A fluorocarbon film is now widely recognized as a cause of the anode effect both in fluoride and in mixed fluoride-containing electrolytes.

In particular, the formation of a blocking film was found to be responsible for the anode effect in mixed chloride–fluoride melts like $MCl–MF–M_2SiF_6$ [22]. The kinetics of gas evolution at the anode and its quantitative composition have been studied in [23]. In addition, the impact of natural hydrocarbon gas on the properties of the electrode under the anode effect was investigated [24].

Relatively to a $KCl–KF–K_2SiF_6$ melt as an electrolyte, following basic facts have been established.

Even a small content of oxides (<1 %) in the electrolyte results in formation of carbon mono- and dioxide as main anode products. After the oxide ions are worked out, the anode gas consists of chlorofluorocarbons of methane series (CF_4, CF_3Cl and CF_2Cl_2) with small admixtures of ethane and ethylene derivatives. The evolution of chlorine gas was not observed in a wide range of melt composition and other experimental conditions. Presumably, the current densities required for the chlorine evolution are higher than the critical values for the onset of the anode effect and, thus, cannot be attained.

The anode reactions include the stages of joint Cl^- and F^- discharge, formation of a solid chlorofluorocarbon $(CF_xCl_y)_s$ on the anode surface and decomposition of this surface compound with liberation of gaseous chlorofluorocarbons:

$$\begin{aligned} &C + xF^- + yCl^- \rightarrow (CF_xCl_y)_s + (x+y)e^- \\ &(CF_xCl_y)_s \xrightarrow{k} CF_nCl_m \uparrow + C \end{aligned} \tag{4.1}$$

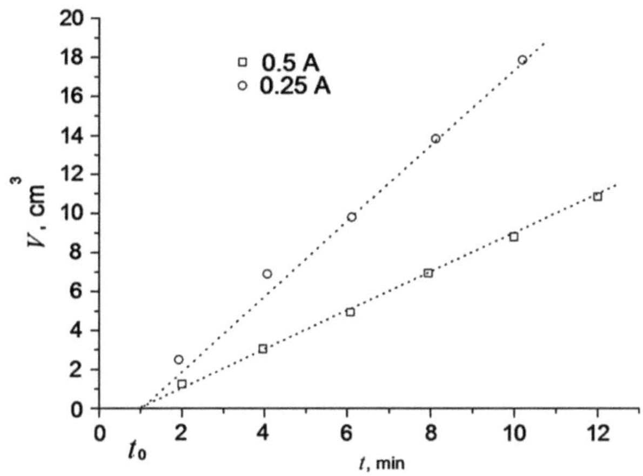

Fig. 4.8 Initial parts of time dependencies for the volume of gas liberated on a graphite anode at the electrolysis of a KCl–KF–K_2SiF_6 melt in galvanostatic conditions for two different current values; temperature 700 °C [22]

The experimental results on the kinetics of the gas evolution supported this mechanism [23]. According to the reactions (4.1), the theoretical time dependency of the volume of the evolved gas can be written as

$$V = \frac{E_V I}{F} t - \frac{E_V I}{kF}\left(1 - e^{-kt}\right) \tag{4.2}$$

where E_V is the volumetric electrochemical equivalent, I is the current and k is the rate constant of the chemical reaction (4.1). Taking the time t sufficiently long, we can neglect the exponential term. This results in the equation of a straight line, which does not pass through the origin of coordinate system:

$$V = \frac{E_V I}{F}\left(t - \frac{1}{k}\right) \tag{4.3}$$

Indeed, the experimental plots $V(t)$ obey such regularity (see Fig. 4.8 as an example). The lines intercept the values of $t_0 = 1/k$ on the time axis, which are, first, practically independent of current and, second, decay exponentially with rising temperature (see Fig. 4.9). Such experimental findings are consistent with the film formation mechanism.

The activation energy calculated from the plots like in Fig. 4.9, 66 kJ mol^{-1}, is in good agreement with reference data [25] on the activation energy of thermal decomposition of fluorinated graphite, 44–61 kJ mol^{-1}.

The stationary state in the accepted model is established in course of "normal" electrolysis with the surface being partly free and partly occupied by the solid carbon halides. The occupied part of the surface is proportional to the applied

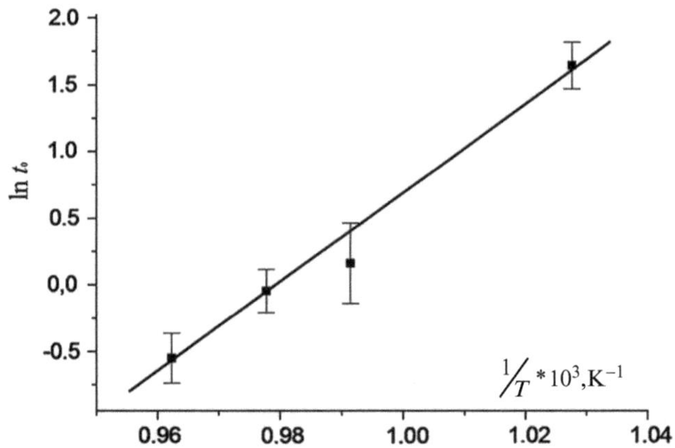

Fig. 4.9 Temperature dependence of the parameter t_0, min (Fig. 4.8) of Eq. (4.3) in Arrhenius coordinates

current density. At some critical value of the current density, the solid film covers the whole surface and the anode effect sets on.

A response of the blocking film upon the exposure to a natural hydrocarbon gas was investigated in [24]. The diagram in Fig. 4.10 shows the changes in the total resistance of the cell before (OA) and after (AB) the input of the hydrocarbon gas on the electrode. The resistance is rising gradually as the anode effect is developing in time and the gas is off. After the gas is on (point A), the resistance is quickly diminishing and, after about an hour, normal conditions of the electrolysis are recovered. As a result, the critical value of current density becomes essentially higher and the anode is no longer consumed because of the oxidation of gaseous hydrocarbons rather than anode material.

The products of this process have been identified by gas chromatography and IR-spectroscopy. Most remarkably, only fluorinated derivatives were found among the gaseous products. The chlorine was bound with hydrogen and evolved from the melt in form of HCl gas.

Following reactions have been suggested for the anode oxidation of the gaseous hydrocarbons in KCl–KF–K_2SiF_6 melt on the surface of a graphite anode:

$$\begin{array}{lr} C_nH_{2n+2} + C_{(\mathrm{graphite})} \rightarrow (C_xH_y)_{\mathrm{solid}} & (\mathit{chemosorption}) \\ (C_xH_y)_{\mathrm{solid}} + mF^- \rightarrow (C_xH_{y-m}F_m)_{\mathrm{solid}} + mH^+ + 2me^- & (\mathit{anode\ oxidation}) \\ H^+ + Cl^- \rightarrow HCl^- & (\mathit{neutralization}) \\ (C_xH_{y-m}F_m)_{\mathrm{solid}} \rightarrow C_nH_{y-m}F_m + (x-n)C & (\mathit{decomposition}) \end{array} \quad (4.4)$$

The behaviour of carbon anodes in molten salts media at the conditions of "normal" electrolysis, being a subject of numerous research works, is understood in general. As for the regime of developed anode effect, the mechanism and dynamics of the processes are much less clear. Only a few works related to cryolite

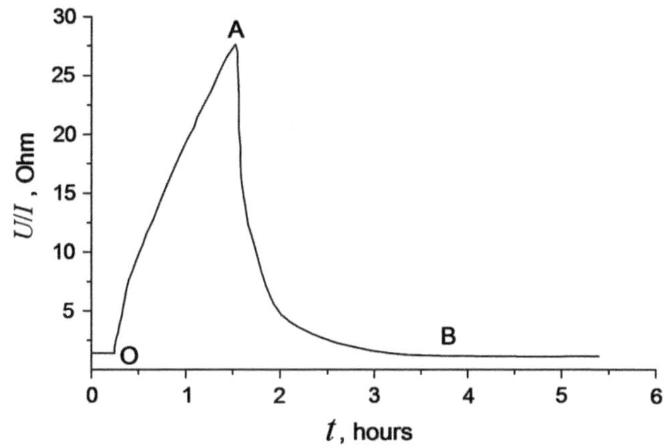

Fig. 4.10 Change in total resistance of the electrolysis cell [24]. The gas flow is on at the time moment, which corresponds to the point A

melts can be mentioned.[3] Some current–potential diagrams were obtained there with sharp current oscillations and very poor reproducibility.

The first complex study of the anode effect dynamics appeared in 1990 [27]. The ideas of dielectric FS and some mathematics of the bifurcation theory were used in this work. We shall consider the results of this study in the next chapter. The end of this section is devoted to some evidence for *cathodic* dielectric FS.

One of the examples, the so-called "cathode effect" in Al electrolysis, has been reported in 1966 [28]. This typical blockade phenomenon occurs at Al cathode at very high current densities of about tens A cm^{-2}. This effect shows itself in sharp increase of ohmic resistance at the cathode followed by arc or glow discharge. The early explanation suggested the formation of a layer of gaseous sodium. In our opinion, formation of dielectric solid film accounts better for this phenomenon.

The formation of a FS with distinct dielectric properties have been found in a system of liquid Al–cryolite melt with additions of silica. This system has been investigated in a series of works [28–31]. Following mechanism has been suggested:

$$\begin{array}{c} (SiO_2)_s \rightarrow Si(IV) \xrightarrow{1\ Al} Si(II) \xrightarrow{2S\ Al} Si(Al) \\ \qquad\qquad\qquad\qquad\quad {\searrow}^{3} \quad {\nearrow}_{4} \\ \qquad\qquad\qquad\qquad\quad [Si(II)]_s \end{array} \tag{4.5}$$

After the silica is dissolved in the melt, fast bulk reaction 1 occurs, where the initial Si (IV) is reduced to the Si(II) intermediate species. The metal aluminium dissolved in the melt[4] acts there as a reductant. Further on, the heterogeneous reaction 2 s takes place at the

[3] See, for example, a relatively recent paper [26] and references therein.

[4] The nature of the dissolved aluminum is still a subject of discussion; intermediate Al(I) and individual Al in form of so-called "metal fog" concepts are in use. Both hypotheses are equally acceptable for the mechanism (4.5).

melt/liquid aluminium interface, resulting in the formation of a liquid Si–Al alloy. This reaction is described in terms of the electrochemical corrosion theory with two conjugated processes: cathodic reduction of silicon (II) and anodic dissolution of aluminium. The rate of cathode half-process is restricted by the limiting diffusion current of silicon species; at 900 °C it corresponds to the rate of silicon reduction [in g(Si) cm^{-2} h] $W_2 = 3.0 \times 10^{-2} C_{SiO_2}$, where C_{SiO_2} is the concentration of silicon in the melt recalculated on wt% SiO_2. The solubility of the intermediate Si(II) is rather low, corresponding to about 0.6 wt% SiO_2. In normal course of the process, the stationary concentration of silicon in the melt is lower than this value and is determined by the rate of supply of the silica to the bath. When this rate would allow being higher than some critical value, the intermediate should deposit onto the surface forming the blocking film. The process 2 then would be hampered and silicon should have to pass into the aluminium through slow reaction 4 between solid substance and liquid aluminium ($W_4 = 0.76 \times 10^{-2}$ g(Si) cm^{-2} h).

It is worthwhile to note that the developed mechanism became a basis for novel process of Al–Si alloys production in industrial Al cells by controlled introduction of silica into the electrolyte [32, 33]. This process was known since the 1930s of the last century [34, 35]; however, numerous attempts of its practical implementation were unsuccessful. Inevitably, the complications arose that accompanied with growth of temperature, evaporation of the electrolyte and losses of fluorides from the bath, slag accumulation. Ultimately, this situation, which was called the "hot run" of the cell, led to the full disorder and termination of the process.

The above mechanism was able to explain easily the cause and give the means to avoid this "hot run". It is clear that silicon can be consumed by the liquid alloy with the rate less than or equal to the rate of the reaction (4.2). This maximum possible rate is equal to about 2×10^{-2} g cm^{-2} h. This number determines the upper possible rate of silica input into the bath that must not be exceeded to avoid the "hot run". Also it is clear that the upper limit exists for the concentration of silicon in the product alloy. For a particular cell, it can be estimated from the formula:

$$C_{Si}^{max} = \frac{i_{Si}E_{Si}}{i_c E_{Al} - i_{Si}(E_{Al} - E_{Si})} \tag{4.6}$$

where $E_{Si} = 7$; $E_{Al} = 9$ are the electrochemical equivalents of silicon and aluminium; i_c is the total cathode current density. According to the above, the maximum current of silicon is $i_{Si} \cong 0.07–0.08$ A cm^{-2}. In most types of industrial cells the total cathode current density is about $i_c \cong 0.5–0.7$ A cm^{-2}. This gives the estimation for maximum concentration of electrochemical Al–Si alloy $C_{Si}^{max} \approx 9 - 12$ wt%. Such alloy can be obtained without technological complications using rigorous monitoring of the flow of silica into the cell.

Before finishing this chapter, let us stop now and take a brief look back. There we can see many complicated experimental phenomena that have got the classification and explanation in the framework of the electrode FS ideas—alas, so far only qualitatively! Remember that we are now very far from simple equilibrium situation; everything is blurred and uncertain here; each phenomenon does not easily give up to a quantitative theory. However, some hopes still remain—the bifurcation theory as a method for investigation of the dynamic behaviour of non-equilibrium systems. We shall try this approach in the final Chap. 5.

References

1. Velikanov AA (1974) Electronic-ionic conductivity of non-metal melts. In: Delimarskii YuK (ed) Ionic melts, vol 2. Naukova Dumka, Kiev, pp 146–178
2. Mott NF, Davis EA (1979) Electronic processes in non-crystalline materials. Clarendon Press, Oxford
3. Reznichenko VA, Ustinov VS, Kariazin NA, Khalimov FB (1983) Chemical technology of titanium. Nauka, Moscow
4. Konstantinov VI (1977) Electrolytic production of tantalum, niobium and its alloys. Metallurgia, Moscow
5. Spitsyn VN, Drobashova TM, Kazanskiy AN (1983) On the alkali tungsten bronzes obtained by electrolysis of molten isopolytungstanates. In: Mokhosoev MV (ed) Chemistry of Mo(VI) and W(VI) compounds. Nauka, Novosibirsk
6. Tchernov RV, Boiko OI, Diubova LD et al. (1983) Method for production of silicon. USSR Patent 1085293, 29 March 1983.
7. Vas'ko AT, Kovach SK (1983) Electrochemistry of refractory metals. Tekhnika, Kiev
8. Andriiko AA, Tchernov RV (1983) Electrodeposition of powdered germanium from oxyfluoride melts. In: Delimarskii YuK (ed) Physical chemistry of ionic melts and solid electrolytes. Naukova Dumka, Kiev, pp 6–60
9. Delimarskii YK, Andriiko AA, Tchernov RV (1979) Ukrainian Chem J 45:1237–1239
10. Boiko OI, Andriiko AA, Yakovlev BV et al (1988) Method for production of silicon. USSR Patent 1462847, 31 March 1988
11. Tchernov RV, Boiko OI, Andriiko AA et al. Method for purification of silicon. USSR Patent 1053530, 8 July 1983
12. Andriiko AA, Panov EV, Gorban′ VA, Parkhomenko NI (1991) Rasplavy (Russian Journal "Melts") 2:123–126
13. Mellors JW, Senderoff S (1966) J Electrochem Soc 113:60–64
14. Dodero M (1939) Bull Soc Chim France 6:309–311
15. Boiko OI, Delimarskiy YK, Tchernov RV (1982) Poroshkovaja Metallurgija (Powder Metall) 6:1–4
16. Gorbachev BV, Spitsyna LG (1982) Physics of semiconductors and metals. Metallurgia, Moscow (in Russian)
17. Boiko OI, Delimarskii YK, Tchernov RV (1985) Ukrainian Chem J 51:385–388
18. Grjotheim K, Krohn C, Malinovsky M, Matiasovsky K, Thonstad J (1977) Aluminium electrolysis: the chemistry of the Hall–Heroult process. Aluminium Verlag, Dusseldorf
19. Watanabe N, Fudjii Y, Yoshisawa S (1963) J Electrochem Soc Japan (Denki Kagaku) 31:131–135
20. Mashovets BP, Aleksandrov GN (1970) On the nature of anode effect. In: French–Soviet symposium on the theory of aluminium electrolysis. Tsvetmetinformatsija, Moscow, pp 138–149.
21. Nakayama T, Ogawa T, Watanabe N (1988) J Fluorine Chem 40:407–410
22. Andriiko AA, Boiko OI, Delimarskii YK, Tchernov RV (1982) Ukrainian Chem J 48: 1186–1191
23. Andriiko AA, Delimarskii YK, Tchernov RV (1984) Ukrainian Chem J 50:1174–1179
24. Delimarskii YK, Andriiko AA, Boiko OI, Bandur VA, Tchernov RV (1985) Ukrainian Chem J 51:52–56
25. Kamarchik P, Margrave JL (1977) J Thermal Anal Calorim 11:259–264

26. Haverkamp RG, Rolseth S, Thonstad J, Gudbrandsen H (2001) In: Light metals 2001, Proceedings of the 130-th TMS annual meeting, New Orleans, 11–15 February 2001, pp 81–488
27. Andriiko AA (1990) Rasplavy (Russian Journal “Melts”) 1:65–74
28. Piontelli R, Mazza B, Pedeferri P (1966) Electrochim Met 1:213, cited by [19]
29. Delimarskii YK, Prutskov DV, Andriiko AA, Tchernov RV (1983) Ukrainian Chem J 49: 738–742
30. Prutskov DV, Andriiko AA, Tchernov RV, Delimarskii YK, Khvalin AP (1983) Ukrainian Chem J 49:845–849
31. Prutskov DV, Pirozhkova VP, Khvalin AP (1983) Ukrainian Chem J 49:1027–1031
32. Prutskov DV, Andriiko AA, Tchernov RV (1987) Tsvetnyje Metally (Russ J Non-Ferrous Met) 2:39–43
33. Prutskov DV, Filippenko AV, Artemenko SA et al (1989) Method for production of aluminium–silicon alloys. USSR Patent 1505076, 1 May 1989; 1600392, 15 June 1990
34. Batashev KN, Zhurin AP (1933) Metallurgy 2:66–70
35. Beliayev AN (1944) Metallurgy of light metals. Metallurgizdat, Moscow (in Russian)

Chapter 5
Dynamics of a Non-equilibrium Electrochemical System

Irreversible phenomena are much more stubborn. However even they tend to enter the general order.

H. Poincarè

5.1 Coupling in Electrochemical Systems: Electrochemical Decomposition of Mixed Conductor

As was shown in Chap. 4, the behaviour of real electrochemical systems on long timescales cannot often be predicted from short-term observations. In terms of the approach considered in 1.6, this happens because of the onset of positive feedbacks destabilising the stationary state of the system. Several possible reasons exist for this event to take place. Among them, most significant, however very rarely considered, is the coupling *between the electrochemical processes at separate electrodes.* To understand how it works, we shall first consider the process of electrochemical decomposition of an electrolyte with mixed ion–electron conductivity.

As follows from the previous chapters, a complex interface Metal/MIEC/Electrolyte (MIEC = mixed ion–electron conductor) appears in many processes related to the electrochemistry of polyvalent metals. The model of MIEC in terms of the concept of polyfunctional conductor (PFC) can be a useful approach to deal with the mechanisms of the processes in such systems. The qualitative classification of EFS has been given based on this approach. Further on, we are going to demonstrate that this concept is useful for quantitative (or at least, semi-quantitative) modelling of macrokinetics (dynamics) of the processes in highly non-equilibrium systems. Before doing this, it is worthwhile to outline some basic ideas related to the MIEC. These considerations will also show some restrictions and approximations that are commonly applied in electrochemical practice and which are no longer valid in such kind of systems.

A.A. Andriiko et al., *Many-electron Electrochemical Processes*, Monographs in Electrochemistry, DOI 10.1007/978-3-642-35770-1_5,

The processes in ionic conductors and on the interface between ionic and electronic conductors (metal, semiconductor) are a research subject of classic electrochemistry. Over the past decades researchers paid some attention to the nonclassic electrochemical object—the conductor with the mixed ionic–electronic conductivity where charge is transferred both with electrons (and/or holes) and ions.

Electrolytical production of metals from chalcogenide (in particular, sulphide) compounds was, in fact, the first problem where the researchers faced the essential effect of mixed conductivity in electrochemical practice. Owing to the studies of Velikanov and his team [1–7], we had got the term polyfunctional conductor (PFC) and the main ideas about physico-chemical properties of this object. According to his theory, the electronic conductance of PFC can undergo the semiconductor to metal transformation (Mott transition), which can be detected from the conductivity–temperature dependency. The possibility had been found for the enhancement of ionic conductivity and, thus, for the improvement of electrochemical behaviour of the melt. It was achieved by means of so-called "heteropolar" additives—compounds with ionic chemical bond.

How the system *metal/PFC/metal* behaves when the direct current passes across? To give the answer means to solve the basic problem of the electrochemistry of PFC. The first intuitive idea, of course, was to represent the system by two constant resistances, for electronic and ionic currents, respectively, connected in parallel. It was commonly applied in early studies of the electrolytic decomposition of liquid PFC [1–3]. Unfortunately, the situation turned to be more complex. The analysis given recently shows that the partial currents are the function of the value of current applied [8–10].

The equations for the diffusion and charge transfer processes into the crystalline ionic–electronic conductors were obtained by Wagner [11] and Yokota [12]. They became the basis for study of transport properties of solid electrolytes, in particular, for determination of the electronic conductivity value [13]. However, these theories are no longer true if the Faradaic process of electrochemical decomposition of the PFC occurs at the interfaces. The elementary theory for stationary process at such conditions [8] and some experimental examples [9] are considered below.

5.1.1 The Mathematical Model

If the PFC is connected between two metallic electrodes, Eq. (5.1) for dependence of the electrical current density on the voltage applied to the electrodes has been obtained, in accordance with the Yokota's diffusion theory of electrical conductivity [2]:

$$i_1 = i_\infty \cdot \frac{1 - e^{z_1 \Phi}}{1 + e^{z_1 \Phi}} \tag{5.1}$$

where

$$i_\infty = \frac{\kappa_1}{\delta} \cdot \frac{R \cdot T}{z_1 \cdot F} \tag{5.2}$$

is Wagner's limiting electronic current density.

Equation (5.1) is not true if the Faradaic process of the PFC decomposition takes place. The correct equations (5.3) and (5.4) have been derived [8] describing the situation when both electronic i_1 and ionic i_2 current flows together:

$$i_1 = \frac{(\varphi_\mathrm{d} - \varphi)\frac{\kappa_1}{2\delta}}{1 + \frac{(1 - \mathrm{e}^{-z_1 - \Phi_d})(\mathrm{e}^{z_1 - \Phi_d} - \mathrm{e}^{z_1 - \Phi})}{z_1(\Phi_\mathrm{d} - \Phi)(1 - \mathrm{e}^{z_1 - \Phi})}} \tag{5.3}$$

$$i_2 = (\varphi_\mathrm{d} - \varphi)\frac{\kappa_2}{2\delta} \tag{5.4}$$

where $\Phi = (F\varphi/RT), \cdots \Phi_\mathrm{d} = (F\varphi_\mathrm{d}/RT)$.

These equations provide the basis for the mathematical analysis of the polarisation curves. Such analysis was carried out to gain some important information on the properties and the behaviour of sulphide melts: ionic and electronic conductivities; the efficiency of electrolytic decomposition and its dependence on electrolysis conditions; the values of the stationary voltage φ_c; and dissolution rate of electrolysis products i_c after the current cut-off.

5.1.2 Experiment and Its Analysis

The polarisation curves for the melts of the systems Tl_2S–TlI and Ag_2S–AgCl were obtained by linear sweep voltammetry [14].

The cathodic polarisation curves of the melts Tl_2S–TlI at 60 mol% of TlJ and at three different temperatures are given in Fig. 5.1.

Linear fit of the initial parts of the curves in Fig. 5.1, which apparently correspond to the electronic current prior to the beginning of electrochemical decomposition, has been performed using Eq. (5.1) expressed in a linearised form

$$y = ax \tag{5.5}$$

where

$$y = \frac{i_1 F}{RT} \tag{5.6}$$

$$x = \frac{1 - \exp\left(z_1 \frac{F\varphi}{RT}\right)}{1 + \exp\left(z_1 \frac{F\varphi}{RT}\right)} \tag{5.7}$$

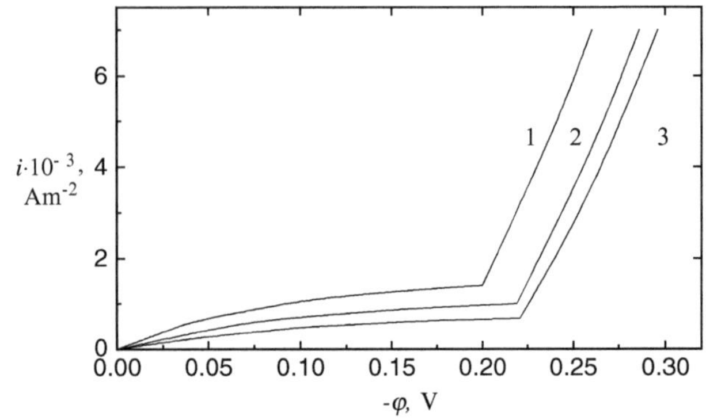

Fig. 5.1 The cathodic polarization curves of melts Tl_2S–TlI at 60 mol% of TlJ and at three temperatures: (1) 788 K, (2) 765 K, (3) 723 K; sweep rate 40 mV s^{-1}. Pt wire was used as a (quasi) reference electrode

$$a = \frac{\kappa_1}{z_1 \delta} \tag{5.8}$$

Figure 5.2 shows the initial parts of Fig. 5.1 in terms of linear equation (5.5).

The values of limiting electronic current densities Eq. (5.2) were calculated from the slopes of the straight lines plotted in Fig. 5.2.

The parameter δ which we can see in Eq. (5.2) can be interpreted as *one-half* of the effective thickness of the sample if the sample is solid and if the stationary distribution of the charge carriers has already been established. In case of a melt where convective mixing in the bulk occurs, the value δ should correspond to the thickness of the diffusive layer where mixing is absent. The exact value of δ in sulphide melts at free convection conditions was not a subject of experimental investigations; the order of value about 10^{-2} cm can be estimated from hydrodynamic considerations.[1] That is why we are not able to extract the precise value of κ_1 from the data on limiting electron currents i_∞.

The second part of the polarisation curves in Fig. 5.1 consists of the sum of current densities i_1 Eq. (5.3) and i_2 Eq. (5.4).

From this part of the curves, the values of decomposition voltage and the ratio of the ionic conductivity to the thickness of the diffusive layer have been obtained by computer numerical calculations.

Figure 5.3 shows an example of such analysis applied to the experimental polarisation curve of the melt Tl_2S–TlI. The calculated curve Eq. (5.4) consists of the partial current densities i_1 and i_2 (curves 2 and 3) described by the Eqs. (5.3) and (5.4); the points correspond to the experimental curve 2 of Fig. 5.1.

[1] Such estimation is not much reliable due to the lack of experimental data on the viscosity and density of the melts at different temperatures. However, it is in agreement with the value $(2 \div 3) \cdot 10^{-2}$ cm which was obtained by direct optical measurements [15] for conditions of free convection in the chloride melts.

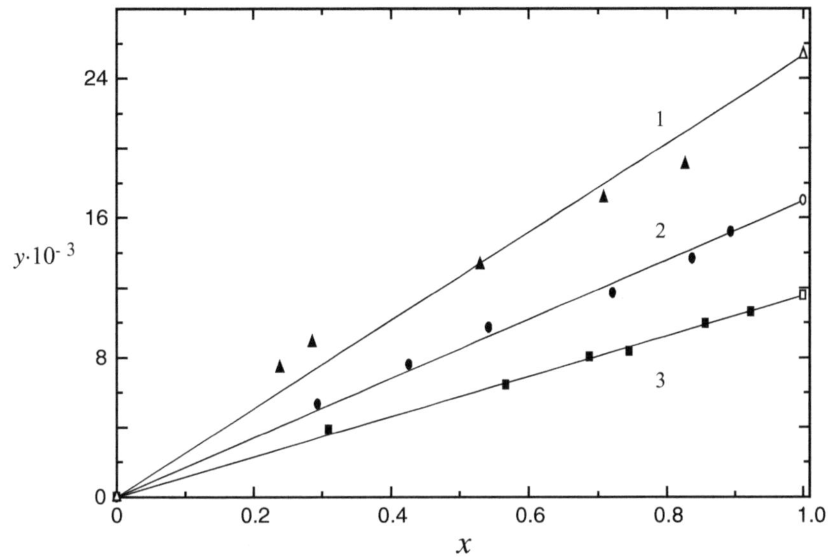

Fig. 5.2 Initial parts of the curves of Fig. 5.1 (*points*) fitted by linear Eq. (5.5) (*lines*); (1) 788 K, (2) 765 K, (3) 723 K

Table 5.1 The results of the analysis of polarisation curves for the system Tl_2S–TlI at 60 mol% of TlJ

T (K)	$i_\infty \cdot 10^{-3}$ (A m^{-2})	$\kappa_1/\delta \cdot 10^{-4}$ (S m^{-2})	κ_1* (S m^{-1})	$\kappa_2/\delta \cdot 10^{-5}$ (S m^{-2})	κ_2* (S m^{-1})	$\delta \cdot 10^4$ (m)	σ_2	σ_2*	$-\varphi_c$ (V)	$i_c \cdot 10^{-3}$ (A m^{-2})	$-\varphi_d$ (V)
723	0.72	1.15	5.3	1.34	62.7	4.61	0.92	0.92	0.212	0.64	0.221
765	1.12	1.70	7.6	1.56	68.4	4.47	0.90	0.90	0.207	0.97	0.219
788	1.72	2.53	10.1	1.81	74.0	3.98	0.88	0.88	0.185	1.41	0.200

Thus calculated, the values of κ_1/δ, κ_2/δ and the ionic conductivity part of the melt are summarised in Table 5.1. The values of the ionic conductivity part and the partial conductivities obtained by independent method are also shown. To the extent of experimental error, the values for the part of ionic conductivity σ_2 obtained from the polarisation curves practically coincide with the data obtained earlier for this system.

The thicknesses of the diffusion layers were calculated from the values κ_1/δ and independently measured κ_1* values; they are also shown in Table 5.1. These data, by order of magnitudes, correspond to the above estimations.

The efficiency of the electrochemical decomposition of Tl_2S η (that is, current efficiency) is determined by the ratio of the ionic current to the total, which, as we can see, is not constant: after reaching the decomposition potential φ_d it tends to the maximal value $\eta_{max} = \sigma_2$ as the current increases. Thus calculated, the values of η are plotted as functions of potential, Fig. 5.4, and current density, Fig. 5.5.

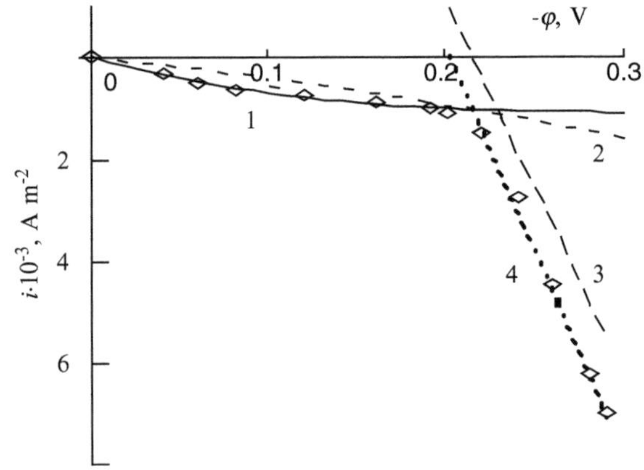

Fig. 5.3 Stationary current–potential dependencies for the melt Tl_2S–TlI at 60 mol% of TlI and $T = 788$ K (*points*). (1) initial part described by Eq. (5.1), (2) partial electronic current density Eq. (5.3), (3) partial ionic current density Eq. (5.4) and (4) total current density

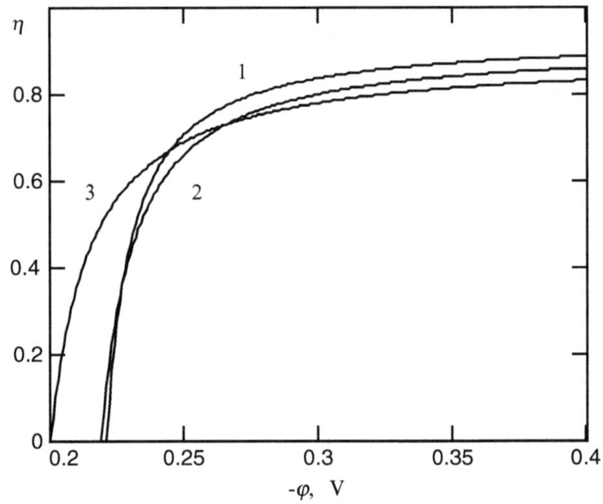

Fig. 5.4 Calculated current efficiencies for the overall reaction $Tl_2S = 2Tl + S$ vs. the potential for the melt Tl_2S–TlI at the temperatures: (1) 723 K, (2) 765 K, (3) 788 K

Similar analysis has been performed with regard to electrochemical decomposition of Ag_2S–AgCl molten mixture according to the overall reaction $Ag_2S = 2Ag + S$. The set of anodic polarisation curves like that shown in Fig. 5.6 was used for the calculations. The results are represented in Table 5.2.

As follows from the obtained data, the contribution of ionic conductivity decreases as the temperature increases. It seems that such regularity is common for all similar systems. The effectiveness of the electrochemical decomposition (the expected current efficiency) essentially depends on the current density of electrolysis and approaches the ionic conductivity part only asymptotically. The calculations show that the current efficiency for the decomposition reaction should smoothly increase with the current density increase. However, in many cases the current density values could be too large to observe such behaviour. In particular, the additional Joule's heat at large current densities should result in the temperature increase and, hence, the decrease in ionic conductivity part; that would result in a current efficiency decrease. Hence, the dependence of the current efficiency on the

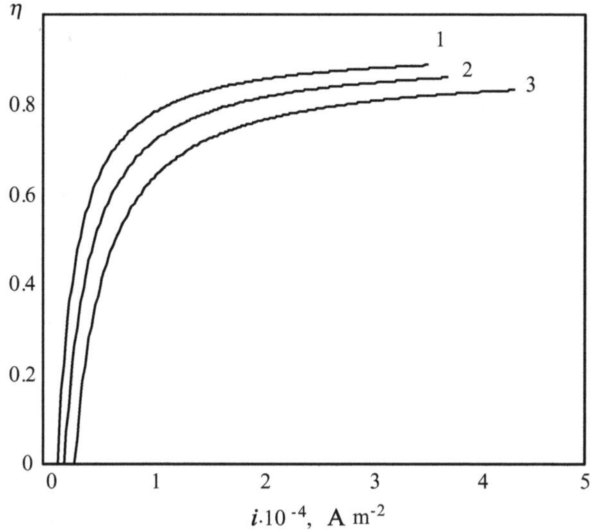

Fig. 5.5 Current efficiencies vs. total current for the same melt at: (1) 723 K, (2) 765 K, (3) 788 K

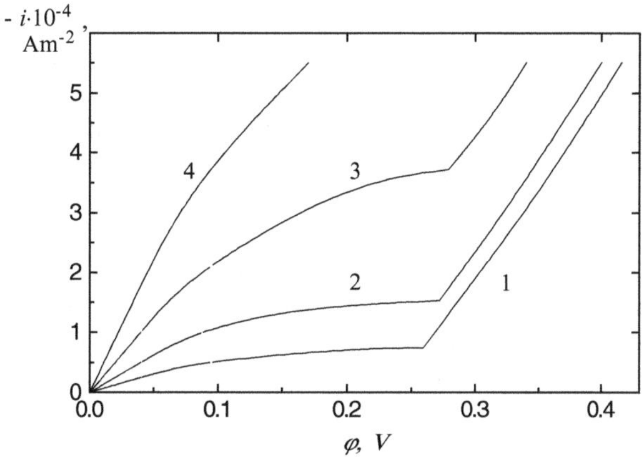

Fig. 5.6 The anodic polarisation curves for the molten Ag_2S–AgCl systems at different temperatures: (1) 673 K, (2) 723 K, (3) 773 K, (4) 823 K, 60 mol% AgCl

Table 5.2 The results of the polarisation curves analysis for the system Ag_2S–AgCl

T (K)	$-i_\infty \cdot 10^{-4}$ (A m^{-2})	$\kappa_1/\delta \cdot 10^{-5}$ (S m^{-2})	κ_1* (S m^{-1})	$\kappa_2/\delta \cdot 10^{-5}$ (S m^{-2})	κ_2* (S m^{-1})	$\delta \cdot 10^3$ (m)	σ_2	σ_2*	φ_c (V)	$-i_c \cdot 10^{-4}$ (A m^{-2})	φ_d (V)
673	0.75	1.30	300	4.03	700	2.3	0.76	0.70	0.229	0.61	0.259
723	1.53	2.46	458	4.46	632	1.9	0.62	0.58	0.223	1.10	0.272
773	3.72	5.76	840	4.62	660	1.5	0.44	0.44	0.188	2.11	0.279
823	6.10	9.44	–	–	–	–	–	–	–	–	–

applied current density is expected to pass through a maximum if the current density varies up to tens A cm^{-2}. This conclusion proves to be true for many cases of the electrolytic decomposition of sulphide melts [16].

The calculated data on the stationary voltages are given in Tables 5.1 and 5.2. After external current is disconnected ($i_1 + i_2 = 0$), the electrolysis products begin to dissolve with the rate $|i_1| = |i_2| = i_c$ at some stationary voltage φ_c. These important parameters were also calculated from the experimental polarisation curves and tabulated. One can see that the rates of dissolution, by the order of a magnitude, correspond to the limiting electronic current, though being a little smaller.

Hence, the attainment of the decomposition voltage is a necessary condition for the electrochemical decomposition of the PFC being possible. This condition is satisfied if the current density i is higher than the limiting electronic current density i_∞. In turn, the value of i_∞ depends on the effective diffusion layer thickness, according to Eq. (5.2). Thus, with other conditions being equal, the decomposition efficiency of a melt should be lower, for example, under forced convection of the electrolyte. Alternatively, the limiting electronic current should be much lower and the decomposition efficiency much higher if the convection is eliminated, for example, by the electrolysis of a porous system filled with an inert matrix or a heterogeneous system where the solid MIEC is distributed inside a liquid ionic conductor. The available experimental data on electrochemical decomposition of the solid suspensions of non-ferrous metals sulphides in molten chlorides [6, 17] maintain this conclusion.

It is important to emphasise that the above theoretical approach for the electrochemical decomposition of PFC has been developed for the *whole* electrochemical system without separating it into the cathode and anode sub-systems. Remembering modern theoretical electrochemistry, we must admit that such approach is not common. Really, the partial cathode and anode processes used to be studied separately at different electrodes (except for corrosion studies). It is believed that the adequate pattern can be obtained for the whole system by "mechanical joining" the separate mechanisms together. Is it valid every time and everywhere? The answer is *no*. We should consider it only as practically useful simplification and remember that there are situations where it is no longer true.

In particular, that is the case for the system considered above. Here the theoretical analysis is hardly ever possible for separate electrode reactions. It becomes clear immediately after we try to define the key parameter, electrode potential, for each single electrode. How can we do it, phenomenologically? How could we determine an electrode reference system and its equilibrium function? How could we define the equilibrium state and how could we determine the deviation from this equilibrium? No way.

Turning back to the Electrochemical Film Systems (EFS), we shall see that a similar situation occurs when the macrokinetics of the electrolysis is investigated. No wonder, the EFS is, in fact, a *one-half* of the whole electrochemical system and must be coupled with the *opposite half-system in any* real process. The correct question is whether or not we can neglect the coupling effect. Given the electrorefinement of polyvalent metal as an example, we cannot do this anymore.

Thus, there are cases when usual electrochemical approach of constructing the whole systems from partial sub-systems gives away. There we step into the uncertain and unexplored world of the "wild" systems governed by unique mechanisms of their own. It seems to be the other step away from equilibrium.

5.2 Coupling in Electrochemical Systems: Electrorefining of Polyvalent Metals in a Long-Term Continuous Process

The simplest model of coupled electrode system (ES) comes up if we connect two electrodes to the bulk of the electrolyte with uniform composition via two adjacent diffusion layers (DL) where the composition of the electrolyte varies:

$$\mathrm{ES_1|DL_1|Electrolyte|DL_2|ES_2} \tag{5.9}$$

In that case, we can approximately estimate the timescale where coupling phenomena could appear by a characteristic time constant τ_c:

$$\tau_c \approx \delta d/D \tag{5.10}$$

where δ is the thickness of the diffusion layers DL, D is the diffusion coefficient, and d is a characteristic dimension of the system, which, most naturally, could be taken as the ratio of the electrolyte volume to the area of the interface. Then, the coupling phenomena can be neglected if the characteristic time Eq. (5.10) is much larger than the timescale of the investigated problem. Otherwise, we could expect for some effects which would not show themselves in short timescale experiments.

Such effects can be especially significant in many modern miniature electrochemical devices like sensors or batteries. For example, in a common Li battery, the spirally wound thin electrodes of metal foil are separated by an electrolyte layer of 10^{-2} cm order of thickness. The diffusion coefficient of Li in the electrode material is about 10^{-8} cm^2 s^{-1}, the path of diffusion being of about 10^{-3} cm. Then the characteristic time is estimated as 10^3 s, that is, less than an hour. Since operation time of a battery (charge or discharge) is several hours, one can suspect that the coupling effects are important. Indeed, here we have the situation familiar to battery researchers—the behaviour of the electroactive materials in standard electrochemical cells with large volume of the electrolyte and in a real battery surrounding is often quite different. Thus, the reliable data on the properties of electroactive battery materials can be obtained only in real batteries or mock-up cells with similar dimensions. That was established empirically and became a common research practice in the recent years.

Traditional electrochemical devices, electrolysis cells, have characteristic dimension of $d \sim 10$ cm, diffusion layer of $\delta \sim 10^{-2}$ cm, and diffusion coefficient about $D \sim 10^{-5}$ cm^2 s^{-1}. That gives the characteristic time $\tau_c \sim 10^4$ s, that is, of 10 h order. Such timescale is important for pilot or industrial electrolysers operating continuously. The dynamics of such long-term process can then be affected by coupling feedbacks. The game of these feedbacks in case of long-term electrorefining of a polyvalent metal is considered below [18].

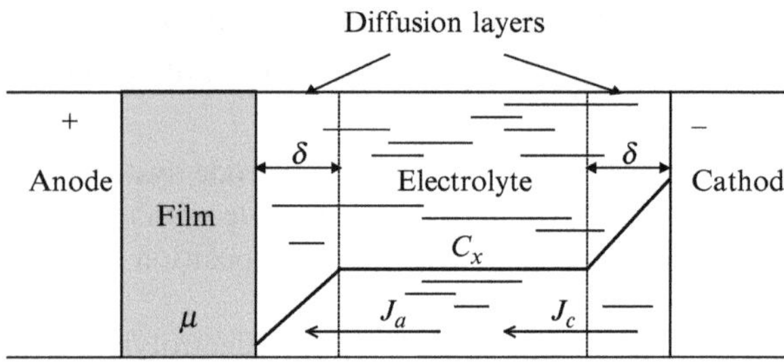

Fig. 5.7 Simplified scheme of an electrorefining system. Distribution of the low-valence compound concentration C_x and fluxes J_c and J_a are shown

The qualitative model of long-term electrorefining process has been considered before (see Sect. 4.1). In that case, the scheme Eq. (5.9) for coupling between cathode and anode systems can be rendered by Fig. 5.7.

This draft represents a simplified model when an anodic film of low-solubility intermediates is typical. The film conductivity is regarded to be both ionic and metallic (see Sect. 4.1). The Faradaic process at the inner junction (that is, metal/film interface) is bound up with the ionic current component giving rise to the film growth. The metallic part of the conductivity causes oxidation of low-valence intermediates at the outer junction (film/electrolyte), both transported from the bulk by the flux J_a and the film's constituents.

Concentration of the intermediate (C_x) in the bulk of the electrolyte and the film thickness (or total amount of film substance) μ are independent variables in the present model. Increasing C_x leads to an increased consumption rate of the intermediate and thus to the increased rate of accumulation of the film deposit. These are, accordingly, the negative and the positive feedback in a dynamic system under consideration.

The quantity of the deposit (denotes here as μ) affects the rate of transport of the intermediate into the bulk of the electrolyte with cathodic flux J_c. The more film deposited, the less is the total bulk concentration of polyvalent compound; the less is the cathodic surface concentration of the intermediate, the lower is the flux J_c and, hence, the lower is the rate of volume accumulation of the intermediate. This influence, therefore, is a negative feedback.

The effect of the amount of the deposit upon its accumulation rate occurs because of the change in the ratio of the parts of conductivity of the film. The most obvious reason for such an effect is an additional heat generation due to Joule heating. This would cause the anode temperature to rise as the deposit is accumulated. Since the temperature dependence of metallic electronic conductivity is weak and ionic conductivity (the F^- ion migration in the cases considered in previous chapter) is of an activation nature, the temperature increase should give rise to an increase in the ionic conductivity part and to an increase in the rate of accumulation of the anodic deposit. Thus, the effect of the amount of deposit μ on its accumulation rate is a positive feedback.

As mentioned in Chap. 1 (Sect. 1.6), such set of intrinsic and crossed feedbacks may result in a variety of possible dynamic behaviours of the system—loss of stability of steady states, oscillations, and other peculiarities. The mathematical analysis given below supports this conclusion.

5.2.1 *Simplest Linear Model*

The above considerations may be represented mathematically by the following set of differential equations obtained from the material balances (see Appendix A):

$$\frac{\mathrm{d}\mu}{\mathrm{d}t} = \mu' = \frac{(\alpha \cdot N - i)}{i \cdot (N - i) \cdot F} + S_{\mathrm{a}} \cdot \frac{D}{\delta} \cdot C_i, \tag{5.11}$$

$$V\frac{\mathrm{d}C_i}{\mathrm{d}t} = V \cdot C_i' = S_{\mathrm{c}}\frac{D}{\delta}k \cdot C_0^{i/N} \cdot \left(1 - \mu\frac{i^2}{N^2 C_0 \cdot V}\right) - S\frac{D}{\delta}C_i \tag{5.12}$$

where I is current, i oxidation number of low-valence intermediate, C concentration, N maximum oxidation number of the polyvalent compound in the electrolyte, S_{a}, S_{c}, S surface areas of anode, cathode, and their sum, V volume of the electrolyte, and α the part of ionic conductivity of the film.

The dimensionless form of Eqs. (5.11) and (5.12) are

$$\frac{\mathrm{d}Y}{\mathrm{d}\tau} = Y' = J + a \cdot X \tag{5.13}$$

$$\frac{\mathrm{d}X}{\mathrm{d}\tau} = X' = b - Y - X \tag{5.14}$$

where

$$\tau = t \cdot \frac{D \cdot S}{\delta \cdot V} \tag{5.15}$$

$$X = \frac{C_i}{k \cdot C_0^{i/N}} \tag{5.16}$$

$$Y = \mu \cdot \frac{i^2 \cdot S_{\mathrm{c}}}{N^2 \cdot C_0 \cdot V \cdot S} \tag{5.17}$$

are the dimensionless variables: time, concentration, and quantity of the deposited film compounds, respectively, and

$$J = \frac{\left(\alpha - \frac{i}{N}\right) \cdot i \cdot \delta \cdot S_{\mathrm{c}} \cdot I}{N \cdot (N - i) \cdot C_0 \cdot S^2 \cdot D \cdot F} \tag{5.18}$$

$$a = \frac{S_c S_a}{S^2} \cdot \left(\frac{i}{N}\right)^2 \cdot k \cdot C_0^{(i-N)/N} \tag{5.19}$$

$$b = S_c / S \tag{5.20}$$

are the dimensionless parameters.

Equations (5.13) and (5.14) represent the simplest linear model of electrorefining. This model, which does not take into account the above-mentioned positive feedback ($\mu' - \mu$), predicts three types of systems behaviour in a long-term process:

1. Unlimited accumulation of the anodic deposit when $J > 0$ (or $\alpha > i/N$).
2. Absence of the anodic film when $|J|/a > b$.
3. Stable steady-state conditions with a limited film thickness when $|J|/a < b$. The transient process may be smooth when $a < 1/4$ or periodic when $a > 1/4$.

5.2.2 *Non-linear Mathematical Model*

The actual behaviour of the electrorefining system can be more complicated owing to non-linear effects, which may manifest themselves in the dependencies of the parameters on the variables of Eqs. (5.13) and (5.14). The influence of the positive feedback between the amount of deposit and its accumulation rate is considered below.

Let us suppose that the parameter J depends on the variable Y because of the temperature dependence of the part of the ionic conductivity of the film, which can be expressed in the form

$$\alpha = \frac{\sigma_i(\theta)}{\sigma_e + \sigma_i(\theta)} \tag{5.21}$$

The temperature dependence of the electronic conductivity σ_e is neglected, and the dependence of the ionic conductivity is assumed to be exponential:

$$\sigma_i = \sigma_0 \cdot e^{\frac{E}{R \cdot T}} = \sigma_{T_0} \cdot e^{\frac{E}{R \cdot T_0} \cdot \theta} \tag{5.22}$$

The dimensionless temperature $\theta = \Delta T / T$ can be expressed as a function of the variable Y from the heat balance with respect to the heat exchange taking place under stationary conditions:

$$q_e = q_t \tag{5.23}$$

The heat evolution (Joule heating) is

$$q_{\mathrm{e}} = \frac{I^2 \cdot \nu \cdot \mu}{\sigma \cdot S_{\mathrm{a}}^2} \tag{5.24}$$

where μ can be expressed in terms of Y from Eq. (5.17).

The heat transfer into the bulk is proportional to the temperature difference

$$q_{\mathrm{t}} = S_{\mathrm{a}} \cdot \beta \cdot \Delta T = S_{\mathrm{a}} \cdot \beta \cdot T \cdot \theta \approx S_{\mathrm{a}} \cdot \beta \cdot T_0 \cdot \theta \tag{5.25}$$

where β is the heat transfer coefficient.

Assuming that a steady-state solution of the Eqs. (5.13) and (5.14) exists (case 3 of the linear model), we can use new variables

$$y = (Y - Y_{\mathrm{s}}) \cdot p; \quad x = (X - X_{\mathrm{s}}) \cdot p \tag{5.26}$$

where

$$X_s = -\frac{J_s}{a};\ Y_s = \sigma + \frac{J_s}{a} \tag{5.27}$$

and, thus, bring Eqs. (5.13) and (5.14) to the normalised form

$$y' = \gamma \cdot f(y) = a \cdot x \tag{5.28}$$

$$x' = -\gamma - x \tag{5.29}$$

The $f(y)$ is a non-linear function of y. Taking into account Eqs. (5.18) and (5.21)–(5.26), it may be written in the form

$$f(y) = \frac{\mathrm{e}^{2 \cdot y} - 1}{\mathrm{e}^{2 \cdot y} + g} \tag{5.30}$$

The dimensionless parameters of Eqs. (5.26), (5.28), and (5.30) are

$$p = \frac{N^2 \cdot C_0 \cdot V \cdot \nu \cdot S \cdot I^2 \cdot E}{2 \cdot i^2 \cdot S_{\mathrm{c}} \cdot S_{\mathrm{a}}^3 \cdot \sigma_{\mathrm{S}} \cdot \beta \cdot T_0^2 \cdot R} \tag{5.31}$$

$$\gamma = \frac{N \cdot \delta \cdot V \cdot \nu \cdot E \cdot I^3}{2 \cdot i \cdot (N - i) \cdot S \cdot S_{\mathrm{a}}^3 \cdot \sigma_{\mathrm{S}} \cdot \beta \cdot T_0^2 \cdot R \cdot D \cdot F} \tag{5.32}$$

$$g = \sigma_{\mathrm{e}} / \sigma_{i,\mathrm{s}} \tag{5.33}$$

The value of the parameter g does not differ significantly from 1. Thus, a hyperbolic tangent can be a fair approximation for the function of Eq. (5.30).

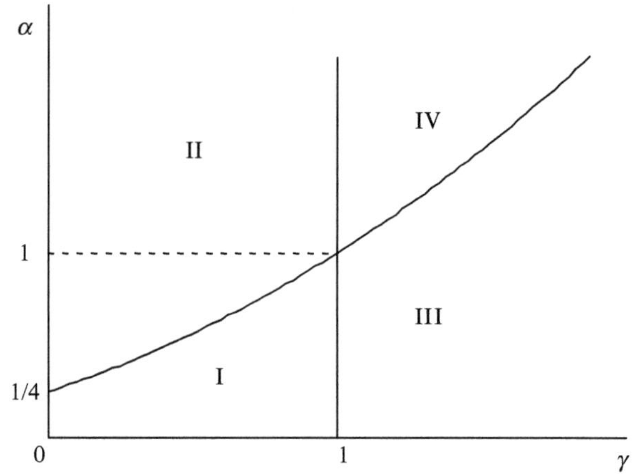

Fig. 5.8 Bifurcation diagram of the system (Eqs. 5.28 and 5.29). Region IV corresponds to oscillatory behaviour

Further, if the system is not far from a steady-state point, the approximate formula for a hyperbolic tangent can be used:

$$f(y) = \frac{e^{2\cdot y} - 1}{e^{2\cdot y} + g} \approx \frac{e^{2\cdot y} - 1}{e^{2\cdot y} + 1} = thy \approx y - \frac{y^3}{3} \tag{5.34}$$

Using the simplifications in Eq. (5.34), the non-linear model Eqs. (5.28) and (5.29) can be reduced to one second-order differential equation:

$$y'' - \left[(\gamma - 1) - \gamma \cdot y^2\right] \cdot y' + \frac{\gamma}{3} \cdot y^3 + (a - \gamma) \cdot y = 0$$

or approximately:

$$y'' - \left[(\gamma - 1) - \gamma \cdot y^2\right] \cdot y' + (a - \gamma) \cdot y = 0 \tag{5.35}$$

Equation (5.35) includes one additional parameter γ; Eq. (5.32). The dependence of the system's behaviour on this parameter is expressed by the bifurcation diagram in Fig. 5.8.

This diagram includes four parts:

1. $\gamma < 1$; $a < (\gamma + 1)^2/4$—stable steady state; a smooth transient process.
2. $\gamma < 1$; $a > (\gamma + 1)^2/4$—stable steady state; damped oscillations in a transient process.
3. $\gamma > 1$; $a < (\gamma + 1)^2/4$—unstable steady state; an unlimited accumulation of the anodic deposit.
4. $\gamma > 1$; $a > (\gamma + 1)^2/4$—unstable steady state; undamped self-oscillations (a limit cycle).

Thus, γ is the bifurcation parameter of the non-linear model. The Hopfs bifurcation occurs at $\gamma = 1$, stability of the steady state being lost and a limit cycle being

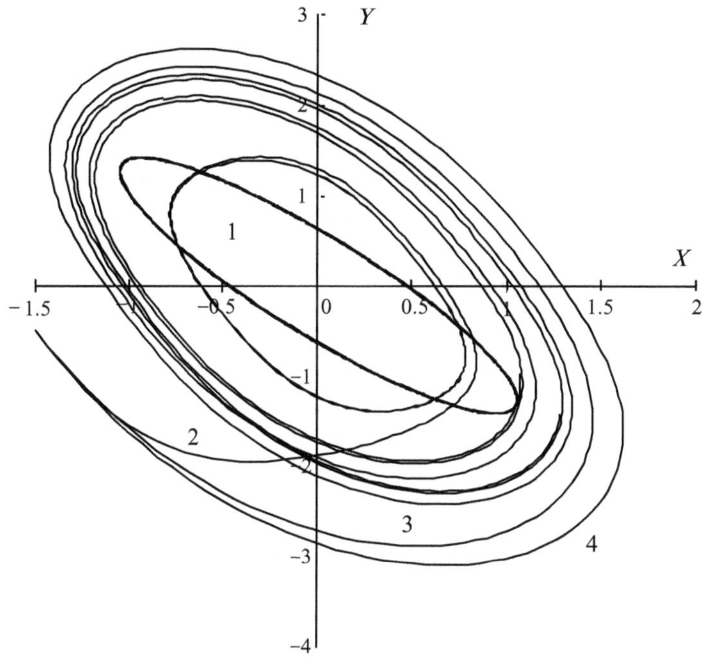

Fig. 5.9 Oscillatory solutions (limit cycles) of the differential equations (Eqs. 5.28 and 5.29) on the phase plane under the following approximations of the non-linear function $f(y)$ according to Eqs. (5.34): (2) $y - y^3/3$; (3) thy; (4) $g = 1.25$ in general formula (5.34). The ellipse (1) corresponds to an approximate harmonic solution by the small-parameter method (see Appendix B); parameter values $a = 3$; $\gamma = 2$

born. The system gains the regime of the undamped oscillations (region 4 in Fig. 5.8). Equation (5.35) corresponds to Van der Pole's equation in this case [19]. Its approximate periodic solution can be obtained by Krylov–Bogoliubov method (see Appendix B) [20]. This solution is represented by the ellipse on the phase plane (Fig. 5.9), where more precise numerical solutions of the system of Eqs. (5.28) and (5.30) are also shown. These have the shape of distorted ellipses in the vicinity of the rough harmonic solution. It confirms the validity of the simplifications of Eq. (5.34).

The approximate formula for the period T of the oscillatory solution in a real timescale is

$$T \approx \frac{2 \cdot \pi \cdot \delta \cdot V}{D \cdot \sqrt{S_{\mathrm{c}} \cdot S_{\mathrm{a}}}} \tag{5.36}$$

Comparing it with a characteristic time (Eq. 5.10) obtained from dimension considerations, we can see that they differ by some constant coefficient close to 2π. Hence, our heuristic considerations as for the effect of coupling were good enough. This important fact allows us for qualitative prediction of possible complicated dynamic situations even when a system is very complex and sophisticated mathematical analysis is not possible.

We are now summarising the results of our theoretical considerations in form of the following conclusions for the electrorefining systems in molten salt electrolytes:

1. As a result of the effects of a solid intermediate film at the anode, long-term continuous processes of polyvalent element electrorefining have features which cannot be predicted from data from short-term laboratory experiments.
2. Three main types of such systems are possible: (a) anodic dissolution without anodic film formation, (b) unlimited accumulation of low-valence compounds at the anode up to complete breakdown of the whole technological process, and (c) the electrolytic process proceeding in the presence of an anodic film of a certain stationary thickness.
3. The presence of additional non-linear effects may cause the loss of the stability of the steady-state condition. There are two possibilities for this case: (a) a change to the conditions of the unlimited accumulation of the anodic deposit, and (b) bifurcation with the birth of a stable limit cycle, the system going over to an undamped oscillation regime.

Turning back to the Chap. 4, we can see that these conclusions are in good agreement with the experimental data available. In particular, the results on the electrorefining of Nb and Ta can be interpreted in terms of the developed theoretical approach as follows. The system under consideration must be related to the type 4 in bifurcation diagram (Fig. 5.8), with an unstable steady state and an oscillatory behaviour. The order of magnitude of the experimental period is in a good agreement with that theoretically predicted. Furthermore, the theory predicts independence (or a very weak dependence) of the parameter $T\sqrt{S_a S_c}/V$ on the operating current and cell design and its inverse proportionality to the diffusion coefficient [see Eq. (5.36)]. This is borne out in practice (see the last column of Table 4.1), the parameter values being $(1.48 \pm 0.11) \times 10^3$ s cm^{-1} (laboratory cell) and $(1.51 \pm 0.18) \times 10^3$ s cm^{-1} (pilot plant cell) for Nb and about twice more, $(3.24 \pm 0.25) \times 10^3$ s cm^{-1} for Ta, whose diffusivity is less.

5.3 Thermokinetic Models of Dynamics in Physico-Chemical Systems: Thermal Self-destruction of Lithium Power Sources with High Energy Density

Previous example demonstrated the upraise of destabilising positive feedback in a coupled electrochemical system due to comparatively week non-linear effect in the governing differential equations (5.28) and (5.29), which resulted from thermal variation in conductivity of the film (Eq. (5.21)). Such non-linearity can be much more expressed in many dynamic systems related to chemical processes in non-isothermal conditions. As will be shown below, a very strong positive feedback can

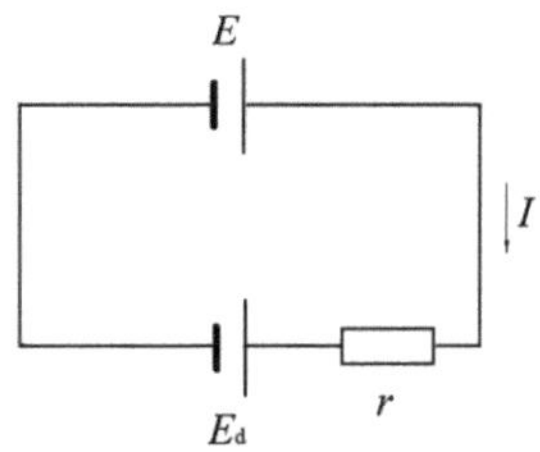

Fig. 5.10 A circuit with a local cell

be borne in case of the impact of an exothermal chemical reaction, the rate of which grows exponentially with temperature according to Arrhenius law. Theoretical models considering this highly non-linear positive feedback in dynamic systems can be defined as *thermokinetic* ones.

To our knowledge, Sal'nikov [21] was the first who developed such type of model and applied it to bifurcation analysis of chemical kinetics in oscillating systems. Further on, the thermokinetic models were successfully applied to the stability problems in the operation of chemical reactors [22]. Lately similar approach was applied to investigation of electrochemical processes [23].

Strong thermokinetic feedback can be significant in many processes related to physical and chemical kinetics. In the model above, this feedback acted implicitly, through the effect of accumulated deposit on the ratio of electronic and ionic conductivity due to the temperature increase. We are now to consider the models where the thermokinetic feedback will act explicitly through the differential equation of heat balance. The temperature in such model acts as one of the independent variables. First, we shall demonstrate the main principles of thermokinetic modelling discussing a relatively simple example—analysis of thermal runaway of highly energy-intensive chemical power sources.

Highly energy-intensive chemical power sources (CPSs), such as lithium thionylchloride systems, are known to be subject to thermal destruction during storage [24]. Thermodynamic causes of this phenomenon have been analysed in [25]. Szpak et al. [26] have shown that the self-destruction can be initiated by an external local heat source. Some cases are known, however, in which the self-destruction proceeds without any external action, i.e. only due to internal causes. These causes, which were analysed in [27], are considered below.

5.3.1 A Model of Chemical Power Source (CPS) Self-Discharge

It is assumed that CPS self-discharge during storage is not uniform in the whole cell, but proceeds at local regions, where the film of lithium has defects which provide access for an oxidiser. Such a region, together with a volume of electrolyte—limited within the pores of the separator, and in the corresponding regions of the cathode, anode, and current collectors—can be represented as a local cell with emf E_d, connected in the circuit by its internal resistance r in parallel to the main source (Fig. 5.10).

At the point of the electrolyte contact, a chemical reaction proceeds with the rate:

$$V\left(\frac{\mathrm{d}C}{\mathrm{d}t}\right)_{\mathrm{chem}} = -kS_{\mathrm{d}}C \tag{5.37}$$

where C (mol cm^{-3}) is the concentration of the oxidiser in the defect region, V (cm^3) the volume of electrolyte in the pore, S_d (cm^2) the area of contact of the electrolyte with lithium in the defect region, and k (cm s^{-1}) the heterogeneous rate constant.

The rate constant is exponentially dependent on the temperature:

$$k = k_\mathrm{o}\mathrm{e}^{-E_\mathrm{c}/RT} \tag{5.38}$$

where E_c (~100 kJ) is the activation energy of the chemical reaction. The case is possible, when the reaction rate depends on the diffusion of oxidiser. Then $k_\mathrm{D} = D/l_\mathrm{D}$ is the diffusion constant, where D and l_D are the diffusion coefficient and length of diffusion path, respectively. As to the activation energy, it is smaller by a factor of 3–4.

Due to this reaction, the local cell in the defect region becomes more discharged in comparison with the main cell. A current I arises charging the defect region:

$$I = \Delta E/r = \Delta E\kappa_\mathrm{S}S \tag{5.39}$$

where $\Delta E = E - E_\mathrm{d}$, κ (Ω^{-1} cm^{-2}) is the conductivity, and S is the area of the defect section which approximates to the section of the separator pore.

E.m.f. decreases due to a decrease in the oxidiser concentration in the pore:

$$\Delta E = \frac{RT}{nF}\ln\frac{C_\mathrm{o}}{C} \tag{5.40}$$

where C_o is the concentration of oxidiser in the initially charged cell.

The rate of charge of the defect region:

$$V\left(\frac{\mathrm{d}C}{\mathrm{d}t}\right)_{\mathrm{el}} = \frac{I}{nF}\Delta E\kappa_\mathrm{S}S = \frac{RT\kappa_\mathrm{S}S}{nF}\ln\frac{C_\mathrm{o}}{C} \tag{5.41}$$

The sum of Eqs. (5.37) and (5.41) is a differential equation of the oxidiser *mass balance* with two variables—concentration and temperature. The conductivity of the cell, κ, is also, in general, temperature-dependent, and this dependence should be similar to Eq. (5.38) if the migration of ions through the film at the lithium surface determines the cell resistance.

The second differential equation presents the *heat balance* of the local cell:

$$W_\mathrm{H} = -W_\mathrm{HT} + W_\mathrm{P} + W_\mathrm{J} + W_\mathrm{C} \tag{5.42}$$

where W_{H} is the heat for heating the local region, W_{P} the latent heat of the reaction (the Peltier heat), W_{J} is the Joule heat, W_{C} is the chemical reaction heat, and W_{HT} is the heat transferred from the defect region.

The heat for heating the defect area W_{H} can be expressed as

$$W_{\mathrm{H}} = Vc_{\mathrm{H}}\rho \frac{\mathrm{d}T}{\mathrm{d}t} \tag{5.43}$$

where c_{H} is the average heat capacity of the local cell and ρ the cell density.

The heat transferred from the defect region is assumed, as usual, to be proportional to the difference in temperature:

$$W_{\mathrm{HT}} = \alpha S_i (T - T_{\mathrm{o}}) \tag{5.44}$$

where α is the effective coefficient of heat transfer, and S_i is the area of heat transfer. The latent heat of the reaction may be conveniently written as

$$W_{\mathrm{P}} = T\Pi = \Delta E \kappa_{\mathrm{S}} S \Pi \tag{5.45}$$

where

$$\Pi = T\frac{\partial E}{\partial T} \tag{5.46}$$

is the Peltier coefficient (V), and $\partial E/\partial T$ is the temperature coefficient of the e.m.f. of the cell. We assume $\Pi > 0$, if the heat is evolved during charge, which corresponds to cells with $\partial E/\partial T > 0$;

the Joule heat:

$$W_{\mathrm{J}} = (\Delta E)^2 \kappa_{\mathrm{S}} S \tag{5.47}$$

and the heat of the chemical reaction in the form:

$$W_{\mathrm{C}} = \Delta H V \left(\frac{\mathrm{d}C}{\mathrm{d}t}\right)_{\mathrm{chem}} \tag{5.48}$$

Substituting Eqs. (5.43)–(5.48) into Eq. (5.42) and introducing dimensionless variables—concentration (the degree of charge):

$$\xi = \frac{C}{C_0} \tag{5.49}$$

and temperature:

$$\Theta = \frac{T - T_0}{T} \approx \frac{T - T_0}{T_0} \tag{5.50}$$

we obtain the final model of CPS self-discharge through the defects in the form of a set of non-linear differential equations:

$$\begin{aligned} \tau_\xi \dot{\xi} &= -\xi e^{\varepsilon_c \Theta} - e^{\varepsilon_r \Theta} K \ln \xi = P(\xi, \Theta) \\ \tau_\Theta \dot{\Theta} &= -\Theta - (p - d \ln \xi) e^{\varepsilon_r \Theta} \ln \xi + H \xi e^{\varepsilon_c \Theta} = Q(\xi, \Theta) \end{aligned} \tag{5.51}$$

where

$$\tau_\xi = \frac{V}{k_0 S_d} \tag{5.52}$$

and

$$\tau_\Theta = \frac{V c_H \rho}{R T_0} \tag{5.53}$$

are constants with the dimensions of time;

$$\varepsilon_c = \frac{E_c}{R T_0} \quad \text{and} \quad \varepsilon_r = \frac{E_r}{R T_0} \tag{5.54}$$

are the dimensionless activation energies of chemical reaction E_c and conductivities E_r (or internal resistance) of the cell, respectively.

The dimensionless parameters of Eq. (5.51), depending on the system properties, are expressed by the following formulas:

$$K = \frac{R T_0 S \kappa_0}{n^2 F^2 S_d C_0 k_0} \tag{5.55}$$

$$p = \frac{R S \Pi \kappa_0}{\alpha S_i n F} \tag{5.56}$$

$$d = \frac{R^2 T_0 S \kappa_S}{n^2 F^2 \alpha S_i} \tag{5.57}$$

$$H = \frac{\Delta H k_0 C_0 S_d}{\alpha S_i T_0} \tag{5.58}$$

5.3.2 *Conditions of CPS Stability During Storage*

Estimation of possible limits of variation of the parameters of the Eqs. (5.51) using data on the properties of the $Li/SOCl_2$ system cells summarised in [27] gave the

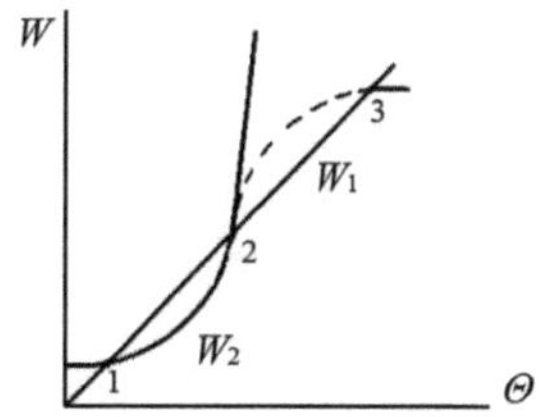

Fig. 5.11 Graphical representation of the critical situation under uncontrolled self-acceleration of an exothermal reaction [28]

following orders of magnitudes: $T_\Xi \sim 10^2$–10^5 s; $T_\Theta \sim 0.01$–1 s; $k \sim 0.1$–100; $p \sim d \sim 10^{-2}$–10^{-5}; $H \sim 10^{-2}$–10^{-4}. Hence, $T_\Xi \gg T_\Theta$, i.e. the time constant for variation of the oxidiser concentration is much less than that for temperature variation. This allows the concentration to be considered as a "slow" variable, and each of Eqs. (5.51) can be analysed separately. Thus, analysing the equation of heat balance we can consider ξ to be constant at any moment of time, whereas the heat exchange proceed under almost stationary conditions:

$$-\Theta - (p - d \ln \xi) e^{\varepsilon_r \Theta} \ln \xi + H \xi e^{\varepsilon \Theta} = Q(\xi, \Theta) = 0 \tag{5.59}$$

In this case, the analysis of the thermal stability of the local cell is similar to that used in the theory of heat explosion [28] (Fig. 5.11). The stationary condition at any moment of time, according to Eq. (5.59), follows from the condition of equality of the rate of heat transfer, presented in Fig. 5.11 by the line $W_1 = \Theta$, and heat evolution W_2 which, in the model accepted, is equal to the sum of the two last terms from Eq. (5.59). In the general case, the system has three stationary states, of which only two are considered within the model accepted (due to the simplification in determining the dimensionless temperature, Eq. (5.50), and the third—high-temperature stationary state—disappears). The "low-temperature" equilibrium state 1 (Fig. 5.11) is a stationary one, and, if it exists, the local "defect" cell will be stable for an unlimited long period of time, provided that the degree of charge remains constant. In the model accepted, the presence of such stable local cells is the cause of the usual low self-discharge of a CPS during storage.

Combinations of parameters are possible, however, at which, during storage, curve W_2 will gradually shift upwards. The stationary states 1 and 2 will converge and, at some critical value of dimensionless variable, ξ_τ will disappear. This bifurcation will lead to the development of local heat source and to catastrophic destruction of the cell. The condition of this is

$$Q(\xi_\tau, \Theta_c) \geq 0 \tag{5.60}$$

where Θ_c is determined from the equation:

$$\frac{\partial Q}{\partial \Theta_{(\Theta=\Theta_c)}} = 0 \tag{5.61}$$

The appearance of instability has been analysed for two limiting cases of Eq. (5.51):

1. The internal resistance of the cell decreases with increasing temperature in accordance with the same regularity as an increase in the rate of chemical interaction, i.e. $\varepsilon = \varepsilon_r$.
2. The internal resistance of the cell is temperature-independent, $\varepsilon_r = 0$. From Eqs. (5.60) and (5.61) functions are found, whose sign determines the cell stability for these two limiting cases:

$$F_1 = e\varepsilon(H\xi - (p - d\ln\xi)\ln\xi) - 1 \tag{5.62}$$

$$F_2 = \ln e\varepsilon\xi - (p - d\ln\xi)\ln\xi \tag{5.63}$$

At $F < 0$ the local defect is stable, whereas at $F > 0$ unstable.

It is obvious that in a freshly charged cell ($\xi = 1$) the defect should be stable, because otherwise it fails immediately after its production. This is possible under the following condition:

$$e\varepsilon H < 1 \tag{5.64}$$

But even if Eq. (5.64) is satisfied, this does not guarantee stability of the cell during a long storage.

The combination of parameters, at which the defect is stable (at some unchanged stationary ξ_{st} value), is determined from the inequality $F(\xi_{st}) < 0$, where the value of ξ_{st} is determined from the set of equations $P(\xi_{st}, \Theta_{st}) = Q(\xi_{st}, \Theta_{st}) = 0$ [see Eqs. (5.51) and (5.61)]. For case (1), this inequality cannot be written in the explicit form, because ξ_{st} should be found numerically from the transcendental equation $\xi_{st} + b\ln\xi_{st} = 0$. For case (2) it has the form

$$KH\ln\varepsilon H + 1 + p - \frac{1}{\varepsilon} + \frac{d}{\varepsilon KH} < 0 \tag{5.65}$$

As the analysis shows, the two parameters K, Eq. (5.55), and H, Eq. (5.58), exert the greatest effect on the local cell stability. This can be represented by a rough diagram, Fig. 5.12, where the line represents the left-hand side of Eq. (5.65), while the last three terms are much less than 1 and, therefore, can be neglected.

The form of the curve in Fig. 5.12, as well as the position and height of maximum depends, to a certain extent, on the value of p and d which display the contribution of the latent heat of reaction and the Joule heat, but the qualitative form of the diagram remains unchanged. The region of stationary stability is limited by the curve and the x-axis, whereas the shaded part of this region corresponds to an absolutely stable defect, where the condition of initial stability Eq. (5.64) holds

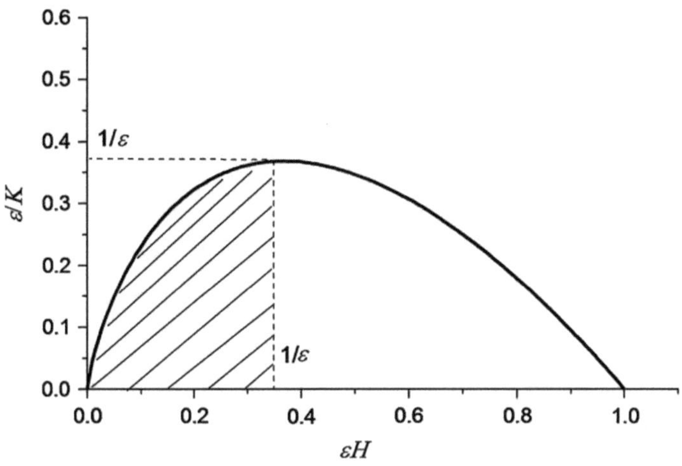

Fig. 5.12 A diagram of mutual influence of K and H on the stability of the local cell

true. With all other combinations of parameters the defect is unstable and, with time, can become a local heat source, capable of initiating thermal destruction of the cell.

A possible development of the process is represented by Fig. 5.13, where numerical solutions of a set of differential equation (5.51) are given at a combination of parameters which does not fall in the stability region (Fig. 5.12). Calculations are presented for two limiting cases, corresponding to stability Eqs. (5.62) and (5.63).

In both cases, the system behaviour has pronounced "catastrophic" character. After a fairly long induction period (the unit of dimensionless time in Fig. 5.13 corresponds to hours or tens of hours of real time) an abrupt increase in the temperature of the local cell proceeds. The "catastrophe" occurs firstly for case (1), when the internal resistance decreases rapidly with increasing temperature.

This theoretical study gives rise to some practical conclusions. Firstly, to produce a stable source, the condition of initial stability of Eq. (5.64) must be achieved. Consequently, the parameter, determined from Eq. (5.58), should be as small as possible. All parameters in Eq. (5.58), except for S_d, are rigidly prescribed either by physico-chemical properties of the system or by standards for production and operation conditions. As to the area of the defect, where the contact of the oxidiser with lithium occurs (S_d), it depends, in the first place, on the presence of impurities in lithium, including mechanical ones. Therefore, the more pure the anode material is, the lesser is the probability of formation of the local source. Secondly, if the condition of Eq. (5.64) is satisfied, then, to get in the stationary stability region, Fig. 5.12, the K constant, determined from Eq. (5.55), should be as large as possible. An increase in purity and the absence of mechanical impurities in lithium also promote this. Besides, the conductivity of local elements, determined as the product of κ and the area of the separator pore section, should be maximum.

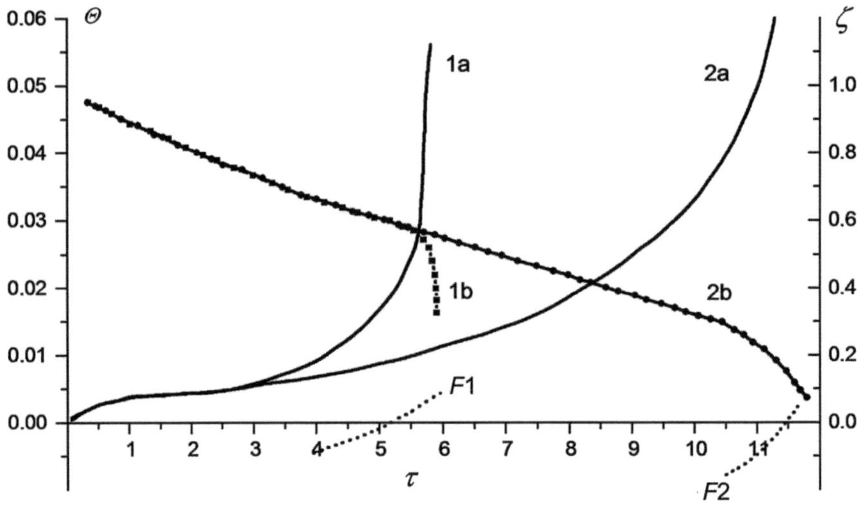

Fig. 5.13 Variation of local temperature (curves *a*) and the degree of charge (curves *b*) of the defect in the approximation of exponential decrease in the internal resistance with temperature (curves 1) and without dependence on temperature (curves 2) as well as variation of stability, Eqs. (5.62) and (5.63) in the critical region. The values of parameters: $\varepsilon = 40$; $H = 0.003$; $d = p = 0.02$; $K = 0.4$; the ratio of time constants $\tau_\xi/\tau_\Theta = 100$; dimensionless time $\tau = t/\sqrt{\tau_\xi \tau_\Theta}$

The probability of formation of the local heat source greatly increases if the distribution of pore sizes in the material of the separator is highly nonuniform and the local defect of the film at lithium falls in an anomalously narrow pore (small *S*). This probability also highly increases in the case of nonuniform electrolyte and the defect falls in a pore, where the concentration of the ionic compound is decreased (small κ). Thus, the appearance of local heat sources, capable of initiating CPS destruction during storage, is less probable; the smaller the content of impurities (including mechanical ones) in lithium, the more uniform the pore sizes in the separator material and the more uniform the distribution of concentration of ionic compound (fluoroborate, lithium preparation, etc.) in the bulk separator.

Thus, the thermokinetic models employ differential equations of material and heat balances when electric current passes through the system. We have just considered the model where the heat equilibrium is fast achieved and the rate of the transient process related to the mass transport is rather slow. This allowed us to simplify the mathematic problem and to extend the analysis as far as to obtain the bifurcation diagram and estimate the time dependence of the variables.

Unfortunately, such simplifications are hardly ever possible in more complicated cases. As an example, we shall consider below the model where only a steady-state stability analysis is possible. Nevertheless, important information can be obtained even in this situation.

5.4 Thermokinetic Model of Electrolysis in the Conditions of Anode Effect

5.4.1 Theory

The qualitative mechanism of the anode effect in fluoride-containing melts has been considered in Chap. 4. The onset of the anode effect gives birth to the surface system of graphite—fluorocarbon film—electrolyte. The film of fluorinated carbon in this system consists of polymer covalent compounds CF_x ($x \approx 1$), where free π-electrons of the original graphite structure are no more available. That is why the conductivity of the film is very low, and the film has pronounced passivative properties.

According to the developed classification (see Chap. 1), this film system belongs to the category of the ion–semiconductor ones with very low total conductivity—that is, to dielectric type of systems. Its low conductivity constitutes of electronic (semiconductive) and ionic (migrative) parts. The electronic band theory prescribes the exponential temperature dependence of the electronic conductivity. Then, the electronic current can be written as

$$i_e = \frac{U}{l}\sigma_s^0 e^{-E_s/RT} \tag{5.66}$$

where U (V) is the potential drop across the film, l (cm) is the film's thickness, and σ_s^0 (Ω^{-1} cm^{-1}) is the pre-exponential coefficient for semiconductivity. The activation energy E_s is determined by the width of forbidden gap, which is rather large for such kind of substances. The electronic (semiconductive) current corresponds to the Faradaic process at the outer surface of the film. This is the evolution of fluorine in purely fluoride melts:

$$2F^- \rightarrow F_2 + 2e^- \tag{5.67}$$

The ionic part of the film conductivity results from the migration of Frenckel's type defects (interstitial ions) in the electric field. It should depend similarly on the temperature:

$$i_m = \frac{U}{l}\sigma_m^0 e^{-E_m/RT} \tag{5.68}$$

where the activation energy E_m is a barrier for an ion to become migratory [29]. The ionic current Eq. (5.68) corresponds to electrochemical fluorination of carbon at the inner surface of the film:

$$C + xF^- \rightarrow CF_x + xe^- \tag{5.69}$$

The fluorocarbon surface film is unstable at elevated temperatures and decomposes to form gaseous fluorocarbons:

$$CF_x \xrightarrow{t} \left(1 - \frac{n}{m}x\right)C + \frac{x}{m}C_nF_m \tag{5.70}$$

The rate of this process, in current units, is

$$i_{Ch} = i_0 e^{-E_{Ch}/RT} \tag{5.71}$$

According to [30], the activation energy E_{Ch} varies from 44 to 61 kJ.

The dynamics of the whole process can be described by the set of differential equations of *material balance*:

$$K\dot{i} = i_m - i_{Ch} \tag{5.72}$$

and *heat balance*:

$$cm(\Delta T) = iU - \alpha\Delta T \tag{5.73}$$

together with the *current balance* equation:

$$i = i_e + i_m \tag{5.74}$$

where i is total current density; partial values of i_e, i_m, and i_{Ch} are determined by Eqs. (5.66), (5.68), and (5.71); c is heat capacity of the anode material; m the mass of the electrode per surface unit; α the heat transfer coefficient for natural convection conditions; K the charge required to form the unit volume of surface substance according to the reaction (5.70); and ΔT the temperature difference between electrode surface and bulk of the electrolyte. The point above a variable denotes, as usual, a time derivative.

In order the system Eqs. (5.72)–(5.74) to be closed, it requires one more equation. This must be an equation of the outer source of electric power, which is usually represented as the dependence of the potential difference on the connected load (electric voltage) on the current passing. The exact form of such equation depends on the particular type of the applied current source. Generally, we can use the equation in the form

$$U = U_0\left(1 - \frac{i}{i_{sh}}\right) \tag{5.75}$$

where U_0 is the open circuit voltage (OCV) of the source, and i_{sh} is the current of the short-circuited source related to the unit surface of the anode. The power source (Eq. 5.75) transforms into a galvanostat if $i = i_{sh}$, and a potentiostat when $i_{sh} \rightarrow \infty$. We shall further use the Eq. (5.75) having in mind that U_0 should be corrected by the e.m.f. value of the electrolysis cell and taking the potential drops in the electrolyte and the current connectors negligibly small relatively to the potential difference across the anodic film.

To simplify the mathematic problem, we assume the equality of the activation energies of electronic Eq. (5.66) and ionic Eq. (5.68) currents (it means independence of the ratio of electronic and ionic conductivities on temperature). Further on, we represent the differential equations (5.72) and (5.73) in dimensionless form using the formulas for particular currents Eqs. (5.66), (5.68), and (5.71) and equation of external source Eq. (5.75):

$$\tau_l \dot{\lambda} = \frac{j v \mathrm{e}^{\frac{\varepsilon \Theta}{\Theta+1}}}{v \mathrm{e}^{\frac{\varepsilon \Theta}{\Theta+1}} + j\lambda} - \mathrm{e}^{\frac{\varepsilon_k \Theta}{\Theta+1}} = P(\lambda, \Theta) \tag{5.76}$$

$$\tau_l \dot{\Theta} = \frac{j^2 v^2 \lambda \mathrm{e}^{\frac{\varepsilon \Theta}{\Theta+1}}}{\left(v \mathrm{e}^{\frac{\varepsilon \Theta}{\Theta+1}} + j\lambda\right)^2} - \Theta = Q(\lambda, \Theta) \tag{5.77}$$

where

$$\lambda = l \frac{i_0^2 \mathrm{e}^{\varepsilon - 2\varepsilon_k}}{\beta^2 \alpha T_0 \sigma} \tag{5.78}$$

and

$$\Theta = \frac{\Delta T}{T_0} \tag{5.79}$$

are dimensionless variables corresponding to the film thickness and temperature difference;

$$\tau_l = k \alpha T_0 \beta^2 \sigma i_0^{-3} \mathrm{e}^{3\varepsilon_k - \varepsilon} \tag{5.80}$$

and

$$\tau_t = \frac{cm}{\alpha} \tag{5.81}$$

are the constants with time units dimension;

$$\varepsilon = \frac{E_e}{RT_0} = \frac{E_m}{RT_0} \tag{5.82}$$

and

$$\varepsilon_k = \frac{E_{Ch}}{RT_0} \tag{5.83}$$

are the dimensionless activation energies of currents and chemical reactions;

$$v = \frac{U_0 i_0 \mathrm{e}^{-\varepsilon_\mathrm{k}}}{\alpha T_0 \beta} \tag{5.84}$$

and

$$j = \frac{i_\mathrm{sh}}{i_0 \mathrm{e}^{-\varepsilon_\mathrm{k}}} \tag{5.85}$$

are the dimensionless OCV and short circuit current of the external source; σ is total conductivity at T_0, and $\beta = \frac{\sigma_\mathrm{m}}{\sigma}$ is the ionic contribution to conductivity.

Thus, we have derived the non-linear mathematic model of the electrochemical process with low-conductive surface film that can decompose with an exponential temperature-dependent rate. Essentially, this is an example of a thermokinetic model.

The model is highly non-linear, and linearization is impossible without loss of the main physical features of the phenomena. Thus, we shall try to analyse the model Eqs. (5.76) and (5.77) as it is, with no simplifications. This is a not so simple mathematic problem, because the model cannot be reduced to any known non-linear differential equation that has already been studied [31].

What analysis is possible, then? First of all, it is the analysis of the stationary solutions (steady states) and their stability, as it was described in Chap. 1.

The stationary values of the variables cannot be written down in terms of explicit functions of the parameters. These have to be determined from the conditions $P(\lambda, \Theta) = Q(\lambda, \Theta) = 0$. The equations below are valid:

$$\Theta_\mathrm{st} = v\left(\exp\frac{\varepsilon_\mathrm{k}\Theta_\mathrm{st}}{\Theta_\mathrm{st}+1}\right)\left(1 - j^{-1}\exp\frac{\varepsilon_\mathrm{k}\Theta_\mathrm{st}}{\Theta_\mathrm{st}+1}\right) \tag{5.86}$$

$$\lambda_\mathrm{st} = v\left(\exp\frac{\varepsilon\Theta_\mathrm{st}}{\Theta_\mathrm{st}+1}\right)\left(\exp\frac{-\varepsilon_\mathrm{k}\Theta_\mathrm{st}}{\Theta_\mathrm{st}+1} - j^{-1}\right) = \Theta_\mathrm{st}\exp\frac{(\varepsilon - 2\varepsilon_\mathrm{k})\Theta_\mathrm{st}}{\Theta_\mathrm{st}+1} \tag{5.87}$$

Following useful equations can be derived from the formulas for stationary state conditions.

First, in the anode effect conditions, the equation of the steady-state polarisation curve can be written explicitly as

$$\Psi_\mathrm{st} = \frac{1}{J_\mathrm{st}}\frac{\ln J_\mathrm{st}}{\varepsilon_\mathrm{k} - \ln J_\mathrm{st}} \tag{5.88}$$

The dimensionless potential drop

$$\Psi = \frac{U}{U_0}v \tag{5.89}$$

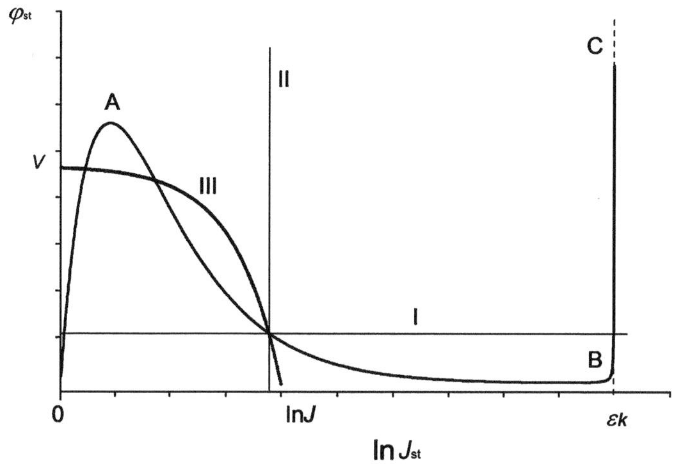

Fig. 5.14 Theoretic stationary polarisation curve calculated from Eq. (5.88) and output characteristics of an external power source: (I) potentiostat; (II) galvanostat, and (III) general case Eq. (5.75)

and dimensionless current

$$J = \frac{i}{i_0} e^{\varepsilon_k} \tag{5.90}$$

are, in steady-state conditions, the functions of the steady-state temperature:

$$J_{st} = \exp \frac{\varepsilon_k \Theta_{st}}{\Theta_{st} + 1} \tag{5.91}$$

and

$$\Psi_{st} = \Theta_{st} \exp \frac{-\varepsilon_k \Theta_{st}}{\Theta_{st} + 1} \tag{5.92}$$

The function Eq. (5.88) gives extremes if $\varepsilon_k > 4$. Taking the above estimations for the activation energy of fluorocarbon decomposition, we obtain $\varepsilon_k \approx 25$–30 [see Eq. (5.82)]. Figure 5.14 represents qualitatively the polarisation curve for such order of ε_k values.

One can see the three different parts in the curve of Fig. 5.14: two ascending (OA and BC) and descending (AB) lines. The ascending branch *BC* approaches asymptotically the critical current value $\ln J_c = \varepsilon_k$, which is the maximum possible in the model under consideration. The temperature then increases infinitely.

The first ascending branch, OA, represents common conditions of the anode effect, when high increase in the applied voltage results in a very small increase in the critical current value. If the current–voltage dependency were taken in

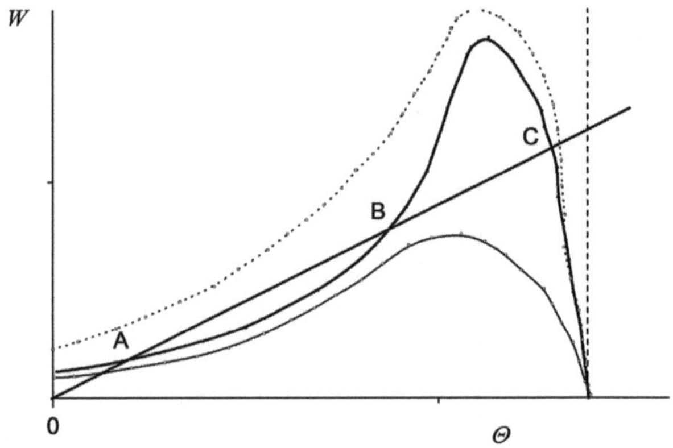

Fig. 5.15 Qualitative plot of heat evolution (*curved*) and heat removal (*straight*) vs. dimensionless temperature Θ

potentiostatic conditions, the system would jump onto the high-temperature branch BC. Temperature should then rise sharply bringing about spark-over the film and glow discharge at the anode, which were observed in numerous experiments; see Refs. in Sect. 4.3.

The most attractive electrolysis conditions belong to the descending branch AB of the polarisation curve. The approximate estimations of currents and voltages in extreme points of the curve in Fig. 5.14 predict possibility of tremendous enhancement of the electrolysis. Since the minimum and maximum currents can differ by 10 and more orders of value, the conclusion follows about intense electrolysis with current densities thousand times higher than initial critical values.

The attainment of such intensive conditions depends upon the shape of polarisation curve $\varphi_{st} - J_{st}$. It is determined by the parameter ε_k—the higher it is the lower and closer to the coordinate origin is the maximum in the curve. Hence, the intensive conditions are the more probable the higher is the activation energy of the chemical decomposition of film compounds. The type of the applied power source is also of great importance. Figure 5.14 shows characteristics of OCV—inner resistance type source (Eq. 5.75) along with its limiting cases: galvanostat ($i = i_{sh}, U_0 \to \infty$) and potentiostat ($i_{sh} \to \infty, U = U_0$).

Hence, the system has 1 or 3 rest points (stationary states). The problem of their stability is the primary task of the analysis.

Preliminary conclusions on the stability of stationary states can be drawn from some qualitative physical considerations. They can be elucidated by means of the plot in Fig. 5.15.

Figure 5.15 presents the temperature dependence of the power of heat generation in the film [which is described by the right side of Eq. (5.86)] together with the sloped straight line corresponding to the intensity of the heat flux into the bulk of the electrolyte [left-hand side of Eq. (5.81)]. Naturally, interceptions of these lines give stationary states (heat generation is equal to heat removal, points A, B, C). Considerations similar to that used in the theory of heat explosion lead to conclusions that the middle point B should be unstable, and the two others (A, C) are stable.

Rigorous analysis of stability was performed by means of Liapunov's theorem, which was written concisely as the inequalities Eq. (1.19). This analysis showed that the initial and end points A and C are stable nodes while the middle B is usually an instable saddle point. This saddle point can sometimes transform into an unstable node or focus, thus allowing for birth of a limit cycle and self-oscillation behaviour. Unexpectedly, the mathematic analysis showed also the possibility of a stable intermediate state in some narrow region of parameters' values (such that the maximum in Fig. 5.15 is not very far from the straight line $W = \Theta$). This result differs from the intuitive physical considerations above.

Now we can summarise the results of theoretical analysis of this non-linear thermokinetic model as follows:

1. In the conditions of an anode effect, the dependency between current and electrode temperature has almost exponential character (Eq. 5.91).
2. The combination of parameters exists where the stable electrolysis at fairly low voltages is possible with current densities of many orders higher than the initial critical values.
3. Apart from the system's properties, the attainment of the intensive regimes depends also on the external power source applied. A potentiostat, as well as common current rectifier with constant voltage output, does not fit because of the instability of "intensive" stationary state. Stability of practical galvanostatic conditions depends on the sign of the derivative $\frac{\partial U}{\partial I}$; ideally, it should be infinite (vertical line II in Fig. 5.14). In practice, it is equal to some large number, which can be both positive and negative, depending on the particular type of power source. If positive ($\partial U/\partial I \to \infty$, which is possible in some electronic regulated sources), the galvanostatic electrolysis should be absolutely unstable. Moreover, galvanostatic devices could hardly ever allow passing over the initial barrier and achieving the descending branch of the polarisation curve. Thus, we can conclude that the only suitable type of current supplier for intensive electrolysis in the conditions of anode effect is the OCV—inner resistance power source with regulated values of output voltage and short circuit current (for example, battery–rheostat).

5.4.2 Experimental Verification

As follows from the model, high activation energy of the chemical reaction is an important prerequisite for "intensive" regime of the electrolysis. Thus, the developed theoretical approach was checked in experiments on the electrolysis of KF–KBF_4 melts with graphite anode [32]. The ability of boron to increase the activation energy of thermal destruction of carbon compounds was taken into consideration when choosing the molten system.

Lead accumulator battery–rheostat was chosen as power supplier. The output voltage and short circuit current were up to 60 V and 50 A accordingly.

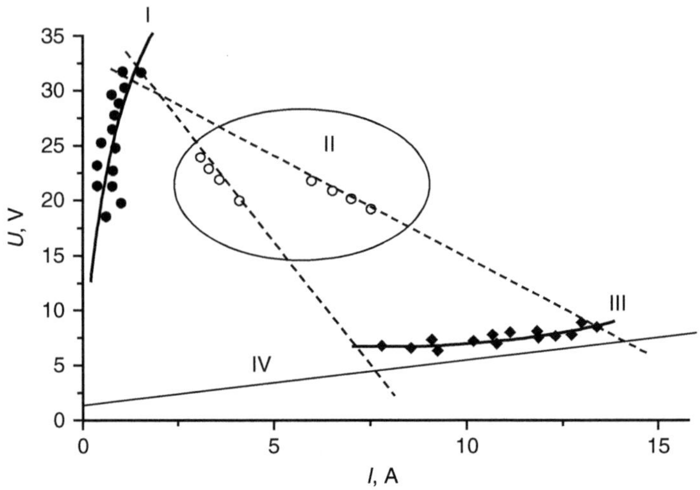

Fig. 5.16 Voltage–current diagram for electrolysis of a KF–KBF_4 melt with periodic small additions of B_2O_3 [32]

The experimental cell was used, offering the possibility to take samples of anode gases for chromatographic and chemical analysis by the method described in [33]. The anode was a graphite cylinder with a thermocouple positioned in a well inside for measurement of temperature. Temperature differences were measured with differential thermocouple, with one junction being inside the anode and the other junction in the bulk electrolyte. Other experimental details can be found elsewhere [32].

Long-term electrolysis of KF–KBF_4 melt was performed at 670 °C with small initial additions (1–2 wt%) of anhydrous boron oxide B_2O_3.

The behaviour of the system in this process is represented in Fig. 5.16 as the plots of operating current vs. cell voltage. Initially, when B_2O_3 is available in the electrolyte, the parameters are fitted by straight line IV, which represents the normal electrolysis with no anode effect complications. The boron oxide is working out of the melt in this process due to evolution of carbon oxides. After it is consumed to practically negligible concentration, the system comes into the conditions of common anode effect jumping onto the line I along the line $U_0 - I_{sh}$, the slope of which depends on the adjusted parameters of the power source. After a new portion of boron oxide is added to the melt, the normal electrolysis on the line IV is recovered, then again the jump follows into the region of low currents (line I), etc.

Further on, the situation is developing in the following way. After 2–3 days of such operation, delays occur in the intermediate region II while the system passes to the conditions of "common" anode effect I. Upon a time, these delays become more and more durable, and then oscillations appear in the region II, as shown in Fig. 5.17. Oscillations vanish after another 3–5 h of the electrolysis, and then, a few hours later, the system jumps into the conditions of "intensive" regime represented by line III in Fig. 5.16. Finally, after 4–5 days operation, the anode

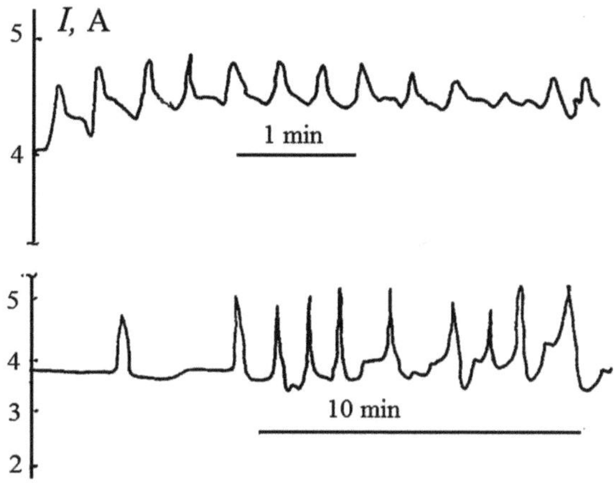

Fig. 5.17 Examples of non-linear self-oscillations in region II of Fig. 5.16

prepared from electrode grade graphite is working stable in region III with no addition of boron oxide to the electrolyte. Thus, stable electrolysis of pure fluoride melt was achieved with 6–10 A current and 8–12 V cell voltage, while the initial critical current at such voltage was less than 20 mA.

Figure 5.15 can explain the transformations of the system as follows.

Initially, the system has one stationary state on the lower dashed curve, which corresponds to low-current region I in Fig. 5.16. Upon a time, the activation energy of chemical decomposition of the surface fluorocarbon film increases due to soaking of the graphite and doping it with boron compounds. The curve in Fig. 5.15 is then shifting upward, which results in the birth of two other new stationary points.

The transition to the situation of the anode effect occurs then as follows. In normal electrolysis conditions (low straight line in Fig. 5.16), the temperature difference between electrode surface and electrolyte bulk is practically zero. After the anode effect takes place, the blocking surface film grows up bringing about essential heat evolution at the electrode–electrolyte interface. The temperature increases, and it permits to achieve the middle equilibrium (steady state) point. This point can be stable at a certain set of parameters, such that the maximum in the heat evolution curve (Fig. 5.15) is not too high. This equilibrium point belongs to the region II of Fig. 5.16. Further increase in the activation energy results in the instability of the intermediate steady-state point; then an attractor of a limit cycle type appeared and self-oscillations (Fig. 5.17) occur. Further on, the intermediate region II (Fig. 5.16) becomes absolutely unstable and the system falls into the region of "intensive" electrolysis III.

Other experimental results [32] are also in agreement with the considered model. Thus, the chemical compositions of anode gases were established by means of chromatography and chemical analysis, and free fluorine and fluorocarbons were formed by the reactions (5.67) and (5.70). Noteworthy is the experimental dependence between current and temperature inside the graphite anode that is given in Fig. 5.18. One can see that it is almost exponential, as the Eq. (5.91) predicts.

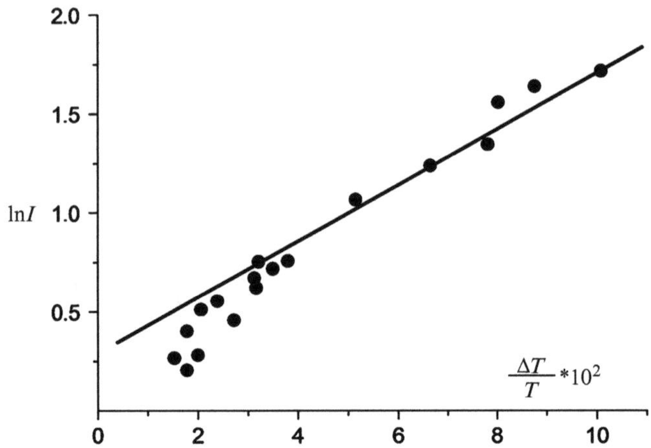

Fig. 5.18 Current–temperature plot for electrolysis of a KF–KBF_4 melt in the anode effect conditions

Appendix A. Derivation of the Differential Equations (5.11) and (5.12)

Because of the mixed conductivity of the anodic film, the Faradaic process at the anode splits into two, i.e. at the inner junction (metal/film) and the outer one (film/electrolyte).

Let us denote the ionic current part as α. Then, at current I and in a lime interval of dt the quantity of deposit

$$d\mu_{\text{int}} = \frac{\alpha \cdot I}{i \cdot F} dt \tag{1a}$$

is formed at the inner junction.

At the outer junction, $(1 - \alpha)I dt$ Coulombs of electricity are consumed at the same time to oxidise the portion of the film substance $d\mu_{\text{ext}}$, and the portion of the intermediate transferred from the bulk by the flux J_c is

$$(1 - \alpha) \cdot I \cdot dt = -(N - i) \cdot F \cdot d\mu_{\text{ext}} + S_a \cdot J_a \cdot dt \tag{2a}$$

To determine the flux, Nernst model of stationary diffusion with linear distribution of the concentration across the layers of the constant thickness is used, and the concentration of the intermediate is assumed to be zero, i.e. the anode discharge proceeds at limit current condition. Then

$$J_a = \frac{D}{\delta} \cdot C_i \tag{3a}$$

The differential equation (Eq. 5.11) is obtained by combination of the equation for the material balance of the film substance

$$\mu' = \frac{d\mu}{dt} = \frac{d\mu_{int}}{dt} + \frac{d\mu_{ext}}{dt} \tag{4a}$$

with Eqs. (1a)–(3a).

Equation (5.12) can be obtained from the balance of the intermediate in the electrolyte, which is supported by the cathodic flux J_c to the bulk and the anodic flux J_a from the bulk

$$V \cdot C_i = S_c \cdot J_c - S_a \cdot J_a \tag{5a}$$

Apart from the above-mentioned Nernst diffusion, the following assumptions and approximations were made in the derivation of Eq. (5.12).

Intervalent equilibrium between polyvalent species at the cathode surface is assumed to establish. A convenient notation of this equilibrium, according Chap. 2, is

$$\left(1 - \frac{i}{N}\right) \cdot E(0) + \frac{i}{N} \cdot E(N) = E(i), \tag{6a}$$

where the symbols $E(0)$, $E(i)$, and $E(N)$ correspond to polyvalent metal compounds, oxidation numbers being zero, i, and N, respectively.

The activity of $E(0)$ is regarded to be 1 (solid pure metal). Further, the concentration C_N of the high-valence compound $E(N)$ at the surface of the cathode and in the bulk is assumed to be equal. This is acceptable for not very dilute solutions. We can now express the cathodic surface concentration of the intermediate by

$$C_{i,cath} = k \cdot C_N^{i/N} \tag{7a}$$

Further, the charge balance equation in the form

$$NC_0 = N \cdot C_N + i \cdot C_i + \frac{i \cdot \mu}{V} \tag{8a}$$

is used, where C_o is the total initial concentration of the electrolyte. This means that the electrolyte at the initial time condition contains only the high-valence compound $E(N)$ with a concentration of C_o, and no additions of salts to the melt during the process are made.

Now the cathodic concentration $C_{i,cath}$ required for the determination of the flux

$$J_c = \frac{D}{\delta} \cdot (C_{i,cath} - C_i) \tag{9a}$$

can be obtained by determining C_N from Eq. (8a) and substituting it into Eq. (7a). The obtained power function may be expanded into a series and, after neglecting the second and higher order members, may be written in the form

$$C_{i,\mathrm{cath}} = k \cdot C_0^{i/N} \cdot \left(1 - \frac{\mu \cdot i^2}{N^2 \cdot C_0 \cdot V}\right) \tag{10a}$$

This linearising simplification is sufficiently correct when the condition is fulfilled

$$C_i << \frac{\mu}{V} << \frac{N}{i} C_0 \tag{11a}$$

Finally, Eq. (5.12) is obtained by substitution of Eqs. (3a), (9a), and (10a) into Eq. (5a).

Appendix B. Approximate Solution of the Non-linear Second-Order Differential Equation by Small-Parameter (Krylov–Bogoliubov) Method [20]

If the non-linear term of the differential equation may be thought of as a small disturbance, the above-mentioned method can then be applied to derive an approximate solution. Equation (5.35) is presented as follows:

$$y'' + w^2 \cdot y + \gamma \cdot f(y, y') = 0 \tag{12a}$$

where

$$w^2 = a - \gamma \tag{13a}$$

and non-linear function

$$f(y, y') = -\gamma \cdot y' \cdot \left(\frac{\gamma - 1}{\gamma} - y^2\right) \tag{14a}$$

The solution is sought in the form

$$y = r(\tau) \cos \varphi(\tau) \tag{15a}$$

by a common technique described elsewhere [19, 20].

Finally, the approximate periodic solution of the limit cycle type is obtained in the form

$$y_{\mathrm{L}} = 2 \cdot \sqrt{\frac{\gamma - 1}{\gamma}} \cos(w \cdot \tau + \varphi_0) \tag{16a}$$

where the dimensionless frequency w is approximately equal to $\sqrt{a-\gamma}$. The dimensionless period $T = 2\pi/w$ of the oscillation, then, is equal to $2\pi/\sqrt{a-\gamma}$. Therefore, the formula of Eq. (5.36) for the real timescale period is accurate to a constant coefficient. This coefficient is of the order 1 and should increase as the current I and concentration C_0 increase. This is hard to verify experimentally because of the very long duration of the tests.

References

1. Velikanov AA (1971) Electrochemical investigation of chalcogenide melts. Dr of Sciences Thesis, Institute of General and Inorganic Chemistry, Kiev
2. Velikanov AA (1974) Electronic-ionic conductivity of non-metal melts. In: Ionic melts, vol 2. Naukova Dumka, Kiev, p 146–154
3. Velikanov AA (1974) Electrochemistry and melts. Nauka, Moscow
4. Belous AN (1978) Studies on the electrode processes at the electrodeposition of thallium, lead and tin from sulfide and sulfide–chloride melts. PhD (Kandidate) Thesis, Institute of General and Inorganic Chemistry, Kiev
5. Zagorovskii GM (1979) Electrochemical investigations of molten systems based on antimony sulfide. PhD (Kandidate) Thesis, Institute of General and Inorganic Chemistry, Kiev
6. Vlasenko GG (1979) Electrochemical investigations of sulfide–chloride melts. PhD (Kandidate) Thesis, Institute of General and Inorganic Chemistry, Kiev
7. Lysin VI (1985) Studies on the nature of conductivity and electrochemical polarization in Chalcogenide–halogenide melts. PhD (Kandidate) Thesis, Institute of General and Inorganic Chemistry, Kiev
8. Mon'ko AP, Andriiko AA (1999) Ukrainian Chem J 65:111–118
9. Andriiko AA, Mon'ko AP, Lysin VI, Tkalenko AD, Panov EV (1999) Ukrainian Chem J 65: 108–114
10. Andriiko AA, Panov EV, Mon'ko AP, Lysin VI (2002) Visnyk Lvivskoho Universytetu (Bulletin of Lviv University). Chemistry 42:109–113
11. Wagner C (1956) Z Elektrochem 60:4–12
12. Yokota I (1953) J Phys Soc Jpn 8:595–598
13. Hebb MH (1952) J Chem Phys 20:185–189
14. Zinchenko VF, Lisin VI, Il'chenko AI (1991) Ukrainian Chem J 57:389–393
15. Polyakov PV, Isaeva LI, Anokhina VS (1975) Izvestiya VUZov. Tsvetnye metally 6:73–76
16. Shevchuk PP, Velikanov AA, Malinovskii VV, Ivanova EI (1973) Elektrokhimiya 9: 1984–1988
17. Izgaryshev NA, Grigor'ev NK (1936) Zhurnal Obschei Khimii (Russ J Gen Chem) 6: 1676–1683
18. Andriiko AA, Panov EV, Mon'ko AP (1998) J Solid State Electrochem 2:198–203
19. Migulin VV (1986) Fundamentals of oscillation theory. Nauka, Moscow (in Russian)
20. Bogoliubov NN, Mitropol'skiy YA (1974) Asymptotic methods in theory of non-linear oscillations. Nauka, Moscow (in Russian)
21. Sal'nikov IE (1949) Russ J Phys Chem 23:258–270
22. Vol'ter BV, Sal'nikov IE (1981) Stability of operation regimes of chemical reactors. Khimija, Moscow (in Russian)
23. Engelgardt GR, Dikusar AI (1986) J Electroanal Chem 207:1–8
24. Godshall NA, Driscoll JR (1984) J Electroanal Chem 131:2291–2297
25. Kuz'minskii YV, Andriiko AA, Nyrkova LI (1993) J Power Sources 46:29–38
26. Szpak S, Gabriel CJ, Driscoll JR (1987) Electrochim Acta 32:239–246

27. Andriiko AA, Kuz'minskii YV (1993) J Power Sources 45:303–310
28. Semenov NN (1986) Chain reactions, Part I. Nauka, Moscow (in Russian)
29. Tchebotin VN (1982) Physical chemistry of solids. Khimija, Moscow (in Russian)
30. Kamarchik P, Margrave JL (1977) J Therm Anal Calorim 11:259–264
31. Reissig P, Sansone H, Conti P (1974) Qualitative theory of nonlinear differential equations. Nauka, Moscow
32. Andriiko AA (1990) Rasplavy (Russian Journal "Melts") 1:65–73
33. Andriiko AA, Delimarskii YK, Tchernov RV (1984) Ukrainian Chem J 50:1174–1179

Chapter 6
Electrochemistry of Ti(IV) in Ionic Liquids

Each generalization assumes, to some extent, the belief in unity and simplicity of Nature.

H. Poincaré

Above we have considered the mechanisms of many-electron electrochemical processes, which were developed with respect to the electrochemistry of molten salts. In this chapter we will consider more recent data on the many-electron processes in *ionic liquids* (ILs), newly developed electrolyte media which fit perfectly to the developed patterns.

6.1 What Are Ionic Liquids?

Generally, an "Ionic Liquid" (IL) is any liquid media consisting of ions (salts above the melting point). Thus, in a broad sense, the high-temperature ionic melts, which were considered above, are also ILs. However, with the discovery of salts which melt at temperatures below the boiling point of water, these were specified in the literature [1] as Ionic Liquids. In this chapter we will use that definition.

In spite of the long history (some alkylammonium salts were found to be liquid in the early years of the twentieth century [2]), they gained considerable attention as a promising class of solvents for various applications including electrochemical ones, only at the end of twentieth century.

Applications of the ILs in the electrochemistry of polyvalent metals are based on their ability to dissolve the compounds of these metals and to be stable in a wide range of applied voltages. Thus, the requirements are similar for both ILs and classical high-temperature molten salt electrolytes.

Imidazolium-based Ionic Liquids are of considerable interest for electrochemical applications. The hydrogen of the imidazolium group can be substituted easily and therefore a variety of properties are achievable. Probably the most widely used imidazolium-based ILs are the 1,3-alkyl substituted derivatives, due to their low melting points, relatively low viscosity and high conductivities. Their physical and

A.A. Andriiko et al., *Many-electron Electrochemical Processes*, Monographs in Electrochemistry, DOI 10.1007/978-3-642-35770-1_6,

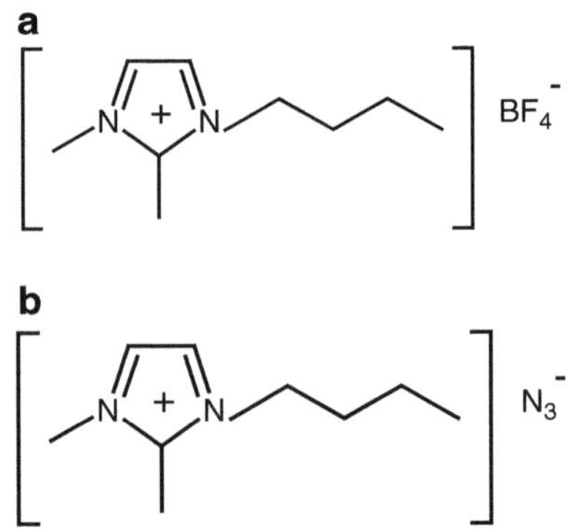

Fig. 6.1 (**a**) Structure of 1-butyl-2,3-dimethyl imidazolium tetrafluoroborate ($BMMImBF_4$) and (**b**) 1-butyl-2,3-dimethyl imidazolium azide ($BMMImN_3$)

some chemical properties have been studied [3, 4]. The relatively low cathodic stability of the disubstituted imidazolium cation limits their application for electrochemical purposes [5, 6]. The cations undergo cathodic dimerisation and dialkylation reactions occur involving the acidic proton in the two-position.

Due to the absence of an acidic proton in the two-position, 1,2,3-trialkylated imidazolium cation-based ILs show superior electrochemical stability compared to the 1,3-substituted ones [7]. Therefore, their use as electrolytes for the deposition of metals is preferable to the disubstituted imidazolium-based ILs.

Electrochemical reduction of Ti(IV) species has been studied in two trisubstituted imidazolium ILs [8–10], both based on the 1-butyl-2,3-dimethylimidazolium cation with BF_4^- ($BMMImBF_4$) and azide ($BMMImN_3$) anions (Fig. 6.1). The solutions of $TiCl_4$ in both salts and TiF_4 in $BMMImBF_4$ were investigated.

Physico-chemical properties, such as conductivity, viscosity and density of $BMMImN_3$, $BMMImBF_4$ and their mixtures, were reported in [11]. Both these salts have similar conductivities and viscosities with the density of the azide IL being about 10 % lower, in the range of the temperatures higher than 65 °C. The electrochemical window of the azide IL is by 1.5 V less than the fluoroborate IL due to the lower anodic stability of the N_3^- anion, though the cathodic decomposition potentials of both ILs vs. Ag/AgCl reference electrode are almost equal.

6.2 Solutions of $TiCl_4$ in 1-Butyl-2,3-Dimethyl Imidazolium Azide

It is known that the products of electrochemical decomposition of fluoroborate salts contain fluorinated compounds [12], which are environmentally harmful. That is why the azide compound, $BMMImN_3$, is considered as a potentially environmental friendlier electrolyte. The electrochemical reduction of $TiCl_4$ solutions in $BMMImN_3$ IL at 65 °C which is about 20 °C higher than the melting point of the IL has been studied in [8]. This salt was synthesised using the reaction of 1-butyl-2,3-dimethyl imidazolium chloride with sodium azide in acetone:

$$BMMImCl + NaN_3 \rightarrow BMMImN_3 + NaCl \downarrow \tag{6.1}$$

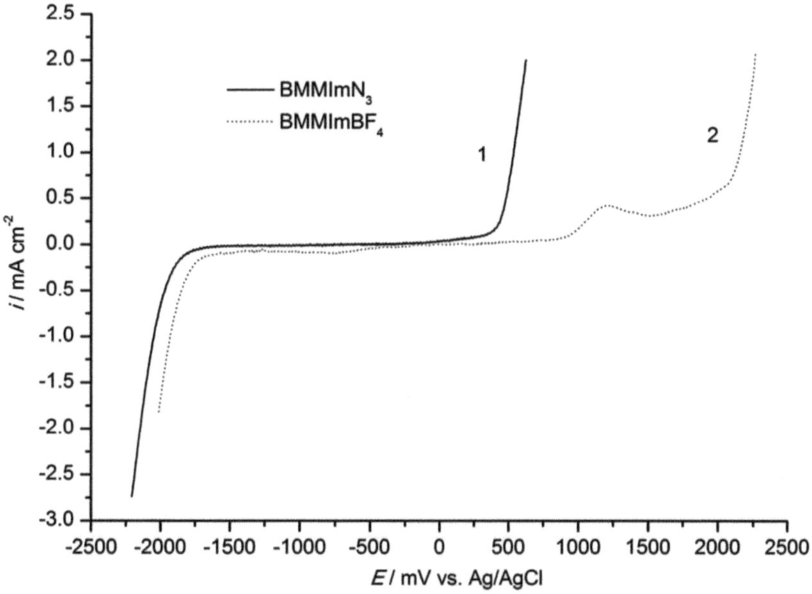

Fig. 6.2 Voltammetry of the pure 1-butyl-2,3-dimethyl imidazolium ionic liquids with (1) azide and (2) fluoroborate cations; potential sweep rate 100 mV s^{-1}, temperature 65 °C, Pt working electrode, Ag/AgCl reference electrode

Linear sweep voltammograms of pure azide and fluoroborate salts are shown in Fig. 6.2. Comparing the electrochemical behaviour of these two Ils, one can see that the anodic stability of $BMMImN_3$ is much lower, whereas the cathodic decomposition potential remains the same. It means that the cathodic limit for the azide salt is also defined by the reduction of the imidazolium-based cation and the decomposition potentials of both ILs are similar.

It was found that $TiCl_4$ has a good solubility in the ionic liquid $BMMImN_3$. Solutions of three different concentrations of $TiCl_4$, i.e., 0.025, 0.05 and 0.12 mol L^{-1}, were studied by CV and chronopotentiometry. Only one broad wave appears in the studied concentration range (Figs. 6.3 and 6.4).

As seen from Fig. 6.2, the wave is much wider than described for a reversible diffusion-controlled process [13].

This, and also a shift of the peak position towards the negative potentials with increasing of the sweep rate, indicates the irreversibility of the reduction process. The peak currents do not obey the Ševčik–Randles equation [13], and the dependence of $(i_p/v^{1/2})$ on the sweep rate, Fig. 6.5, is slightly non-linear, resembling the curves for irreversible process according to Matsuda–Ayabe theory [14].

Figure 6.6 shows an example of a stationary polarization curve for a solution of $TiCl_4$ in $BMMImN_3$. The limiting current value, which is attained at –1 V, is unsteady because of gas evolution starting at a potential of – 1.05 V which can be seen distinctly at the electrode. Thus, one can conclude that the product generated by the reduction

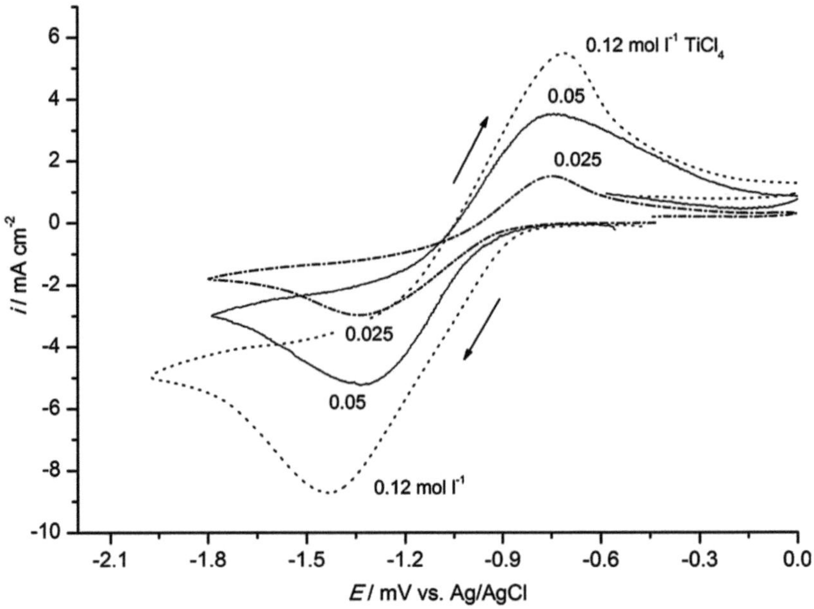

Fig. 6.3 Cyclic voltammograms of $TiCl_4$ solutions in $BMMImN_3$ at 65 °C: (**a**) with different $TiCl_4$ concentrations, potential sweep rate 0.4 V s^{-1}

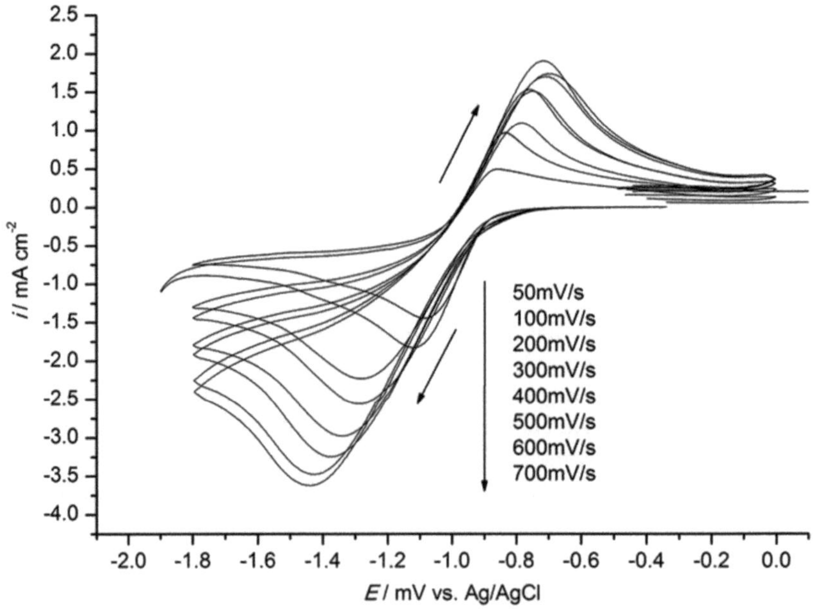

Fig. 6.4 Cyclic voltamograms of $TiCl_4$ solutions in $BMMImN_3$ at 65 °C at different potential sweep rates; concentration of $TiCl_4$ 0.025 mol L^{-1} (potential range 0 to −1.9 V)

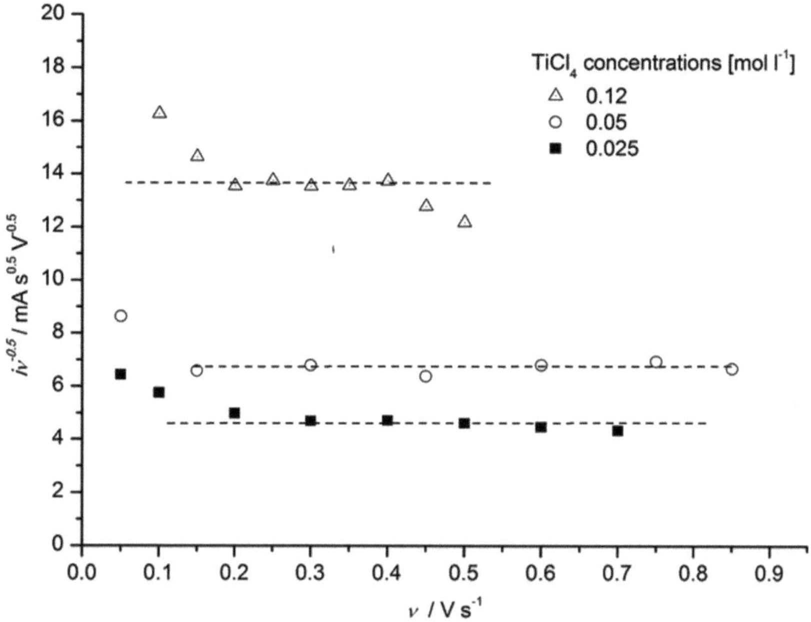

Fig. 6.5 Peak current analysis for cyclic voltammetry studies of different $TiCl_4$ concentrations at 65 °C in $BMMImN_3$

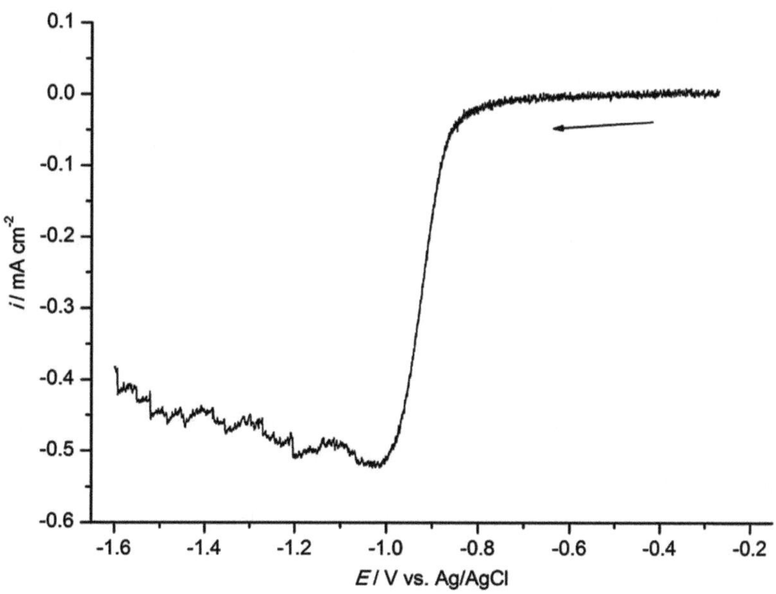

Fig. 6.6 Current–potential curve at a scan rate 1 mV s^{-1}, temperature 65 °C, 0.05 mol L^{-1} concentration of $TiCl_4$

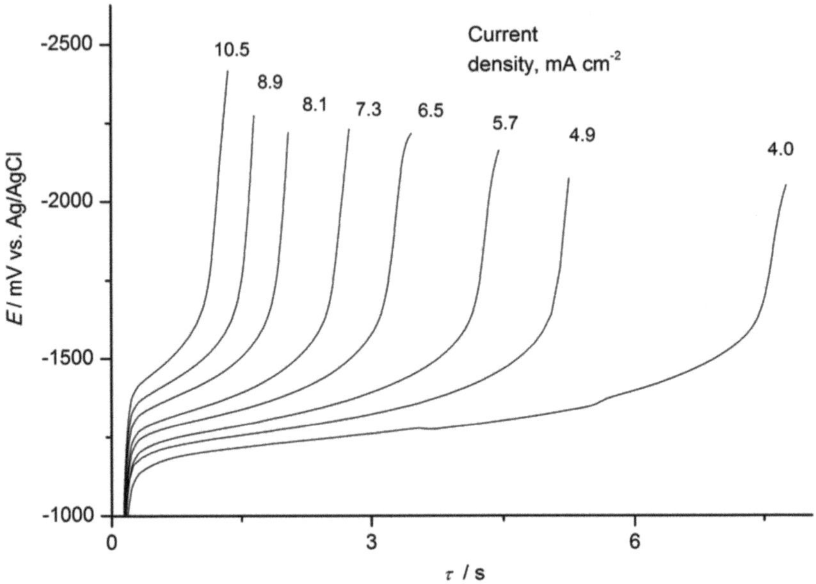

Fig. 6.7 Potential response after applying constant current step 4.0–10.5 mA cm^{-2}; $BMMImN_3$ with 0.12 mol L^{-1}; temperature 65 °C; Pt working electrode

reaction is not stable and gradually decomposes evolving volatile products, since the pure $BMMImN_3$ was found to be stable up to -1.8 V.

Thus, as it follows from voltametric investigation, reduction of Ti(IV) in $BMMImN_3$ electrolyte is a one-step irreversible process. The product of this process, a Ti(III) complex, is not stable and decomposes slowly in the conditions of stationary polarization.

As CV, the potential–time chronopotentiometric curves also contain only one reduction wave (cf. Fig. 6.7).

The shape of the curves is typically non-Nernstian. The plot E vs. $\ln(\sqrt{\tau} - \sqrt{t})$ is linear (Fig. 6.8) in accordance with the theoretical equation for an irreversible process [13]:

$$E = \frac{RT}{\alpha F} \ln\left[\frac{2k_f^\circ}{(\pi D)^{1/2}}\right] + \frac{RT}{\alpha F} \ln[\tau^{1/2} - t^{1/2}] \tag{6.2}$$

The analysis delivers a pre-logarithm coefficient $(RT/\alpha nF) \cong 115$ mV, which obviously corresponds to a one-electron process.

Transition time analysis for different Ti(IV) concentrations is given in Fig. 6.9. The $i\tau^{0.5}$ product is approximately constant and linearly depends on the concentration (Fig. 6.10) in accordance with Sand equation:

$$i\sqrt{\tau} = \frac{1}{2} nFC\sqrt{\pi D} \tag{6.3}$$

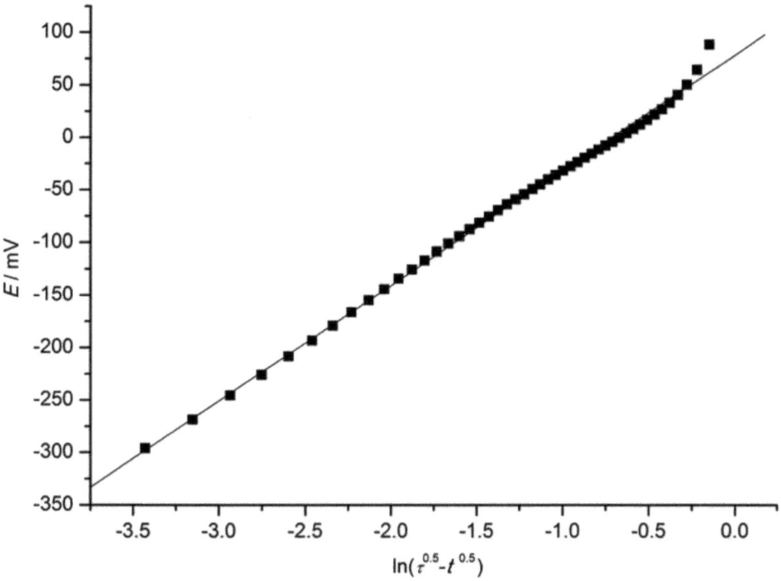

Fig. 6.8 Semi-logarithmic plot of a potential–time curve; $C_{TiCl_4} = 0.12$ mol L^{-1}; current step 4.9 mA cm^{-2}. Temperature 65 °C

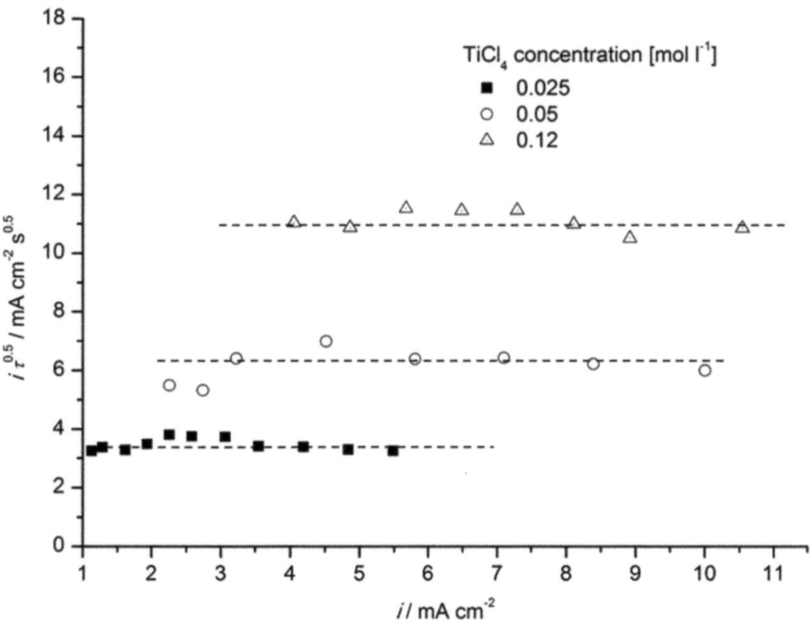

Fig. 6.9 Analysis of the obtained transition times in terms of the Sand equation; $BMMImN_3$ with different $TiCl_4$ concentrations; temperature 65 °C; Pt working electrode

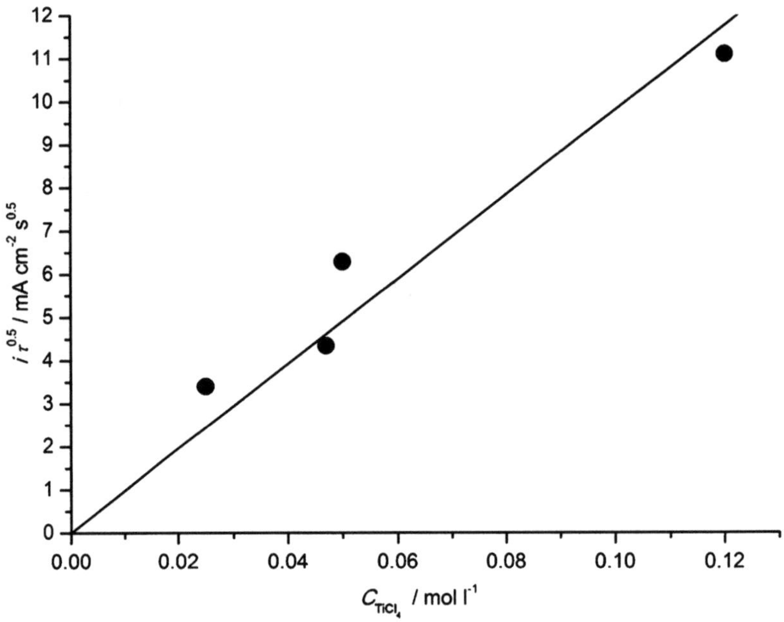

Fig. 6.10 Dependence of $i\sqrt{\tau}$ on the concentration of $TiCl_4$ at 65 °C

where i is the applied current, τ is the transition time and n is the number of electrons involved in the reaction.

Assuming $n = 1$ in Eq. (6.3), from the data of Fig. 6.10, a value of the diffusion coefficient for the Ti(IV) species $(1.3 \pm 0.6) \times 10^{-6}\,\mathrm{cm^2\,s^{-1}}$ was calculated.

Comparison of chronopotentiometry and CV data validates the conclusion that the reduction is a one-electron process. As follows from the theory, the ratios of $(i_p v^{-1/2}/i\tau^{1/2})$ at the same conditions must be equal to

$$\frac{i_p v^{-1/2}}{i\tau^{1/2}} = 0.284\sqrt{\frac{nF}{RT}} = 1.67\sqrt{n} \text{ at } 65\ ^\circ\mathrm{C} \tag{6.4}$$

for a diffusion-controlled reversible process and

$$\frac{i_p v^{-1/2}}{i\tau^{1/2}} = 0.317\sqrt{\frac{\alpha nF}{RT}} = 1.86\sqrt{\alpha n}, \quad \text{also at } 65\ ^\circ\mathrm{C}, \tag{6.5}$$

if the process is totally irreversible.

The average ratio was calculated to be 1.23 ± 0.09, which should correspond to an irreversible process with $n = 1$ and $\alpha = 0.44$.

Thus, the process in the azide electrolyte involves only one electron and is kinetically hindered. As follows from the studies of FTIR spectra [8], the reason for such behaviour is the formation of strong anion complex $[Ti(N_3)_6]^{2-}$, which can be reduced only to Ti(III) azide complex:

$$[Ti(N_3)_6]^{2-} + e^- \rightarrow [Ti(N_3)_6]^{3-} \tag{6.6}$$

The product of this process slowly decomposes with gas evolution.

Thus, the system $TiCl_4$–$BMMImN_3$ cannot be used for deep reduction of Ti(IV) species.

6.3 Solutions of $TiCl_4$ in 1-Butyl-2,3-Dimethyl Imidazolium Tetrafluoroborate

The system $TiCl_4$–$BMMImBF_4$ was studied by CV and chronopotentiometry techniques at 65 °C. The experimental details can be found in [9].

It was found that the pure IL is electrochemically stable in the range of −1.8 to +2 V vs. Ag/AgCl reference electrode (Fig. 6.2). The cathodic limit, as in the case of the azide salt, is caused by reduction of the imidazolium-based cation. In the anodic direction, oxidation of tetrafluoroborate anion, with BF_3 and fluorocarbon formation, is limiting the electrochemical window [12, 15]. The anodic peak at +1.2 V is attributed to the chloride anion oxidation. The impurity of Cl^- results from the synthesis.

The observed pattern differs depending on the concentration. The CV of diluted solutions (Fig. 6.11) shows a first cathodic peak within the range −0.36 to −0.38 V. At concentrations higher than 0.02 mol L^{-1}, an additional cathodic peak appears at +0.03 to −0.05 V. These two cathodic peaks gradually merge with increasing concentration. Up to a concentration of 0.1 mol L^{-1}, a third cathodic peak is observed at −0.9 V (Fig. 6.12).

The dependence of the peak current density on the concentration for the first cathodic peak (*I*) consists of two linear branches (Fig. 6.13). In addition, the peak at −0.9 V vanishes and, instead, a broad irregular cathodic wave appears at approx. −1.25 V.

Thus, the preliminary study of $TiCl_4$–$BMMImBF_4$ system by cyclic voltammetry revealed at least three concentration regions where the electrochemical behaviour of dissolved $TiCl_4$ is essentially different: (1) “diluted”, $C < 0.03$ mol L^{-1}; (2) “medium”, $0.03 < C < 0.14$ mol L^{-1}; and (3) “concentrated”, $C > 0.14$ mol L^{-1}. The reduction processes in these three concentration regions were studied with CV and chronopotentiometry methods.

6.3.1 *Diluted Solution (0.025 mol L^{-1} $TiCl_4$ in $BMMImBF_4$)*

Two cathodic waves are distinctly observed, both in CV and chronopotentiograms (Fig. 6.14), in this concentration region.

The transition times related to the two observed CV peaks at low concentrations are plotted in Fig. 6.15.

The first wave is typically S-shaped if plotted in terms of Karaoglanov equation: $E = f\left(\ln\left(\sqrt{\tau} - \sqrt{t}\right)/\sqrt{t}\right)$ (Fig. 6.16).

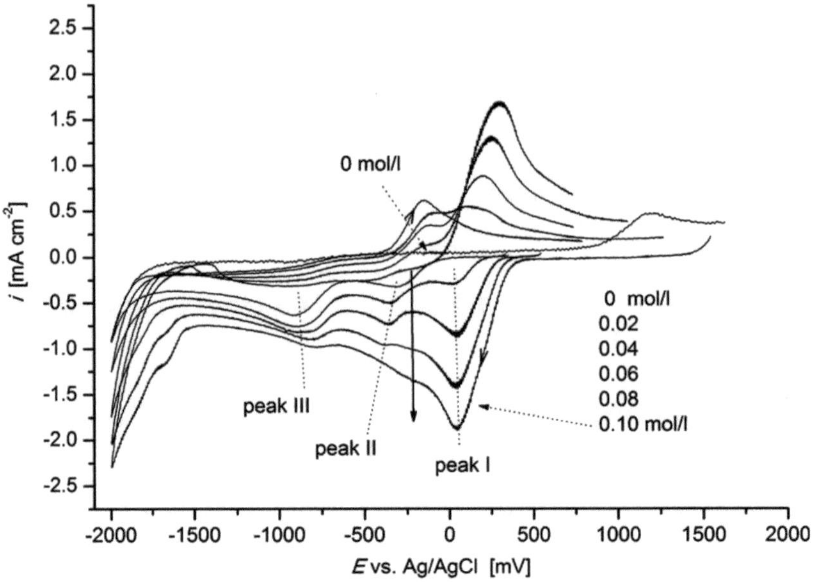

Fig. 6.11 Cyclic voltammograms of $TiCl_4$ solutions in $BMMImBF_4$ at concentrations 0–0.1 mol L^{-1} in concentration steps 0.02 mol L^{-1}. Scan rate 0.1 V s^{-1}

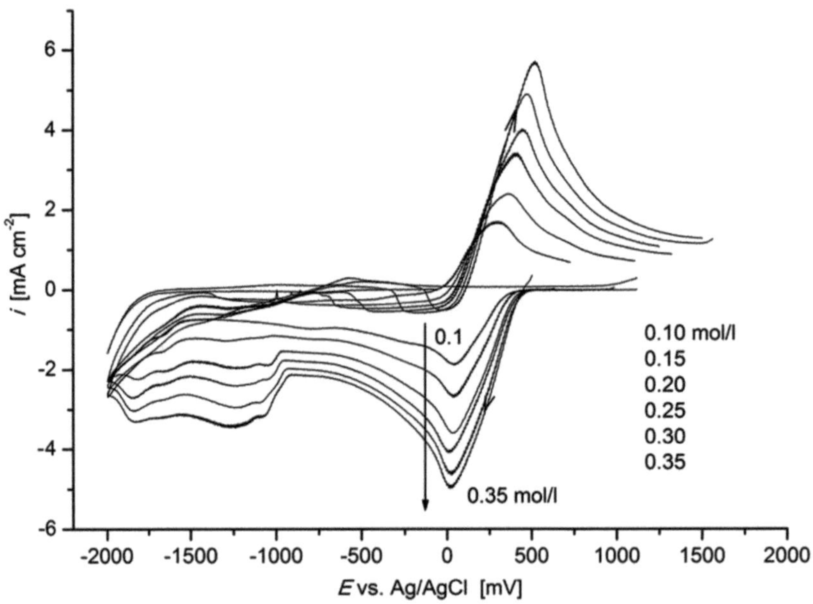

Fig. 6.12 Cyclic voltamograms of $TiCl_4$ solutions in $BMMImBF_4$ at concentrations 0.1–0.35 mol L^{-1} in concentration steps 0.05 mol L^{-1}. Scan rate 0.1 V s^{-1}

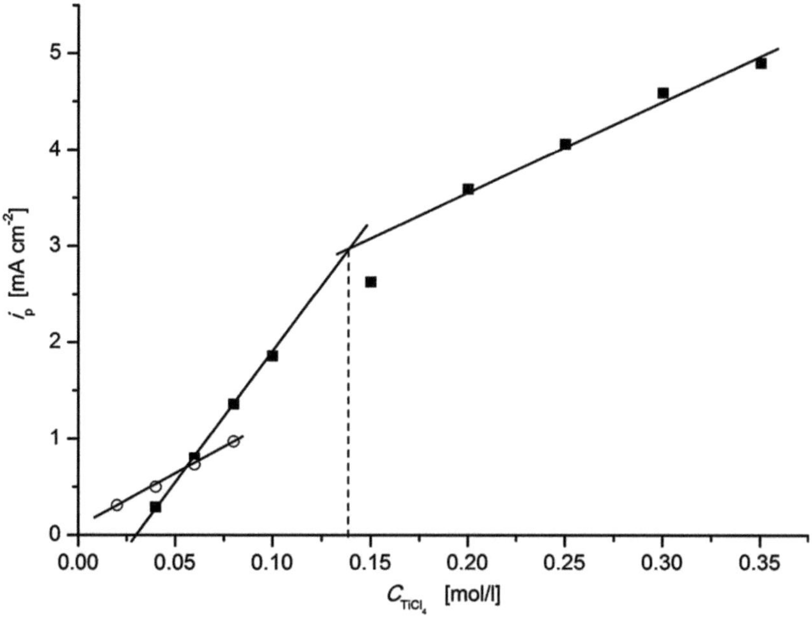

Fig. 6.13 Current of cathodic peak I (*filled square*) and peak III (*open circle*) vs. concentration

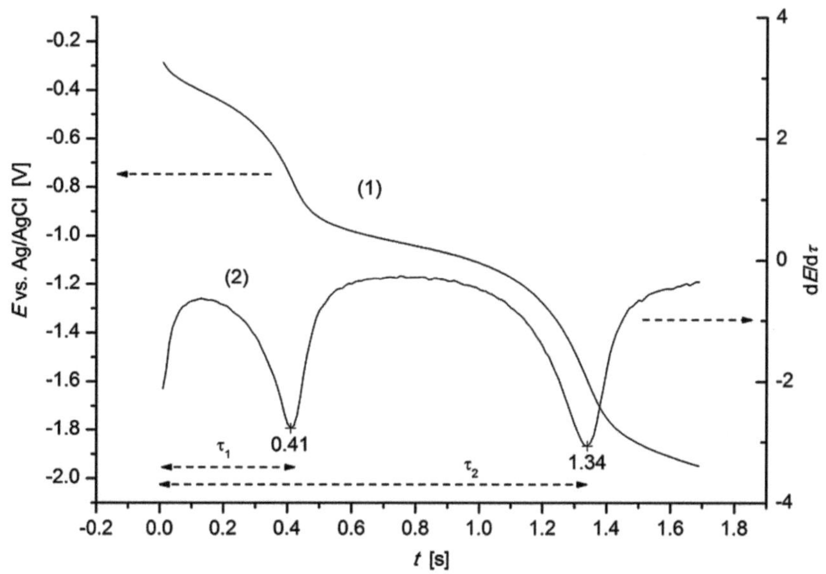

Fig. 6.14 Potential response to a current step -1.45 mA cm^{-2} (curve 1) and its first derivative (curve 2)

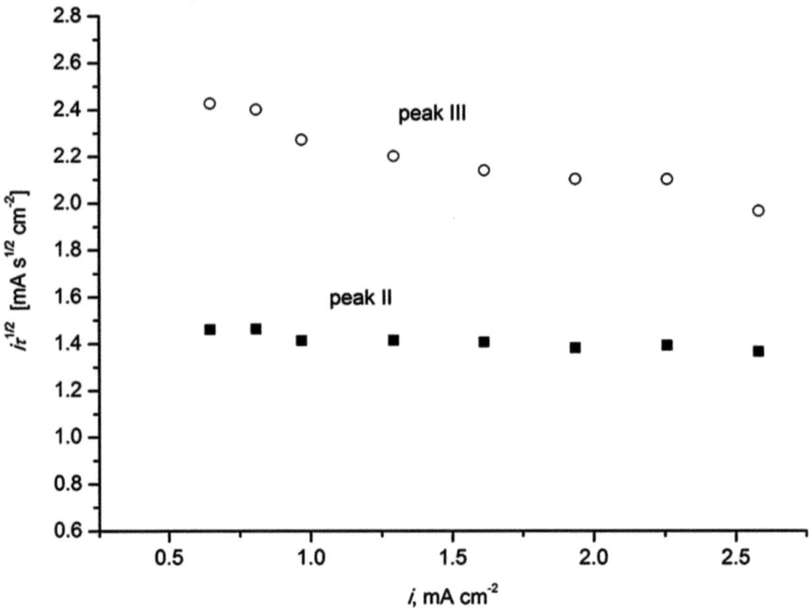

Fig. 6.15 Chronopotentiometric transition time analysis for the waves obtained at 0.025 mol L^{-1} $TiCl_4$ in the electrolyte

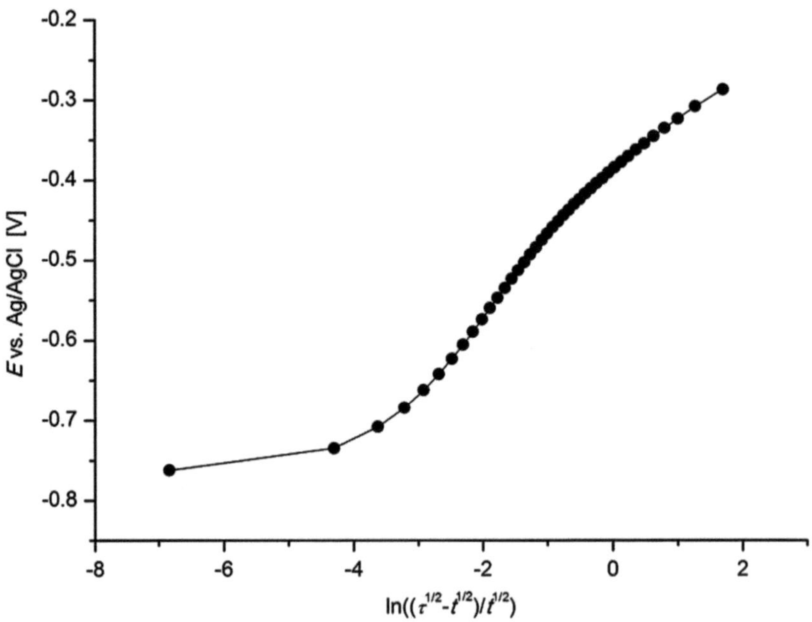

Fig. 6.16 Karaoglanov plot of the potential–time curve in potential range of peak II

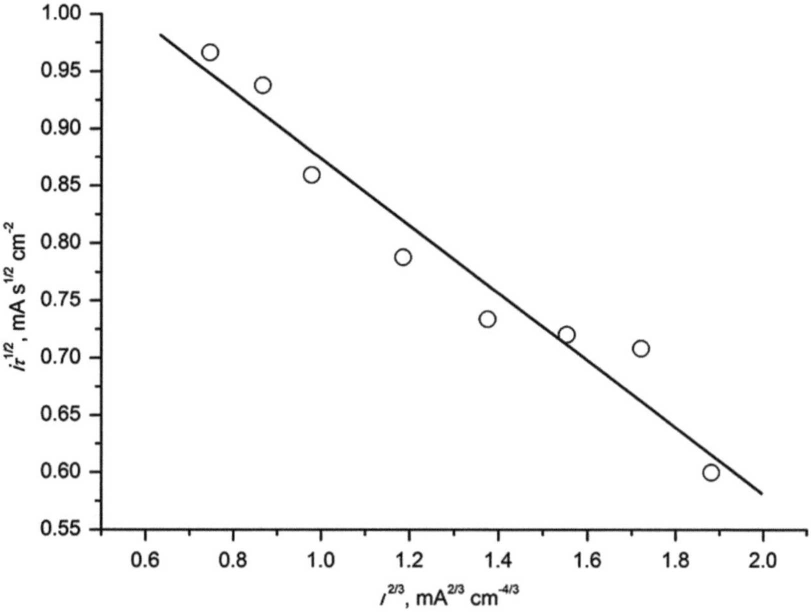

Fig. 6.17 Transition times (Fig. 6.16) analysis in terms of Fischer–Dračka equation [16]

As we have discussed it in Chap. 2, this is evidence of the formation of rather thermodynamically stable intermediate when the reaction of its disproportionation is at equilibrium. Hence the first wave should correspond to the two-electron process: $Ti(IV) \underset{e^-}{\overset{+e^-}{\rightleftarrows}} Ti(III) \underset{e^-}{\overset{+e^-}{\rightleftarrows}} Ti(II)$; $2Ti(III) \rightleftarrows Ti(IV) + Ti(II)$.

The ratio of transition times for second and first waves suggests the formation of Ti(I) as the product of this second process. The negative slope of $i\tau_2^{1/2}$ vs. (i) plot could be caused by its irreversible disproportionation according to the Fisher–Dračka reaction mechanism [16]. To check this assumption, the data were plotted as the function $i\sqrt{\tau} = \left(i\sqrt{\tau_2} - i\sqrt{\tau_1}\right)$ vs. $i^{2/3}$ (Fig. 6.17). It is linear, which allows us to suggest that the process for the second wave can be represented as $Ti(II) \xrightarrow{e^-} Ti(I)$; $2Ti(I) \rightarrow Ti(I) + Ti(0)$.

Fitting the data in Fig. 6.17 with Eq. (3.10), the rate constant of the disproportionation reaction was calculated to be $k_r = (1.95 \pm 0.05) \times 10^6\ cm^3\ mol^{-1}\ s^{-1}$.

6.3.2 "Medium" Concentration Range

Three consecutive separated waves are observed in the "*medium*" concentration range of $TiCl_4$, up to 0.14 mol L^{-1} (Fig. 6.11). The first wave (peak I) appears at more positive potentials (~0 V), corresponding to the formation of the first intermediate, Ti (III), which becomes more stable as the concentration increases (the relative height of the peak is growing). Figure 6.18 shows the potential response to a current step in the electrolyte with 0.05 mol L^{-1} $TiCl_4$.

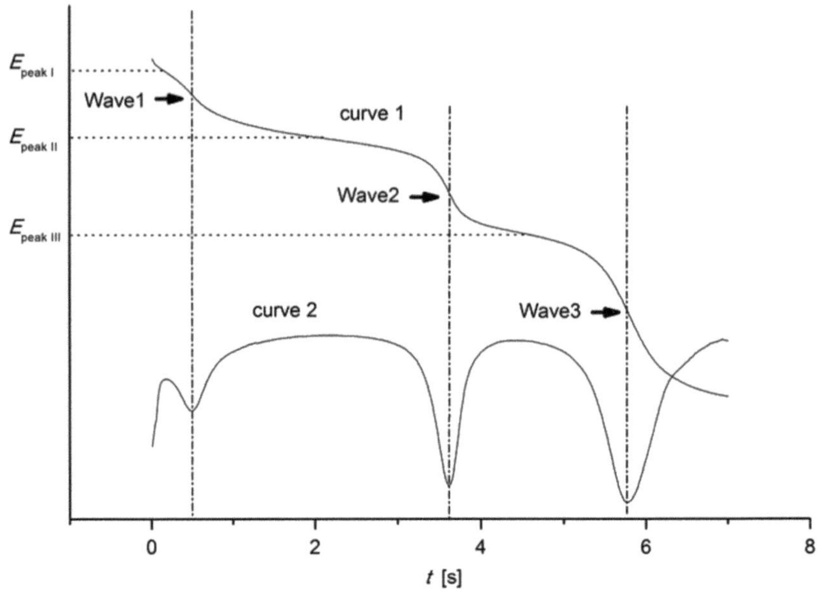

Fig. 6.18 Potential response to a current step −0.97 mA cm^{-2} (curve 1) and its first derivative (curve 2)

The independence of $i\tau^{1/2}$ of the current density is evidence of a diffusion-controlled reaction (Fig. 6.19).

The $i\tau^{1/2}$ ratios in Fig. 6.19 are approximately equal to $i\tau_1^{1/2} : i\tau_2^{1/2} : i\tau_3^{1/2} = 1 : 3 : 4$, which could be attributed to the formation of Ti(III), Ti(I) and Ti(0) reduction products at the end of each reduction step. A similar pattern is observed at a concentration 0.1 mol L^{-1}.

6.3.3 "Concentrated" Solutions

Further increase of the concentration results in a phase separation in the melt: small drops of undissolved $TiCl_4$ are clearly visible on the surface and in the bulk of the electrolyte.

Accordingly, the reduction pattern becomes much more complicated at potentials more negative –1 V mol L^{-1}. Series of poorly resolved waves appear in potential region about −1 to −2 V in the CVs (Fig. 6.12) and up to five peaks are observed in the derivative curve of the CP data (Fig. 6.20).

The CP transition times and CV peaks for concentrated solutions can be hardly interpreted in terms of simple stepwise reduction mechanism. The reason of such behaviour can be related to the phase separation, which is observed in "concentrated" solutions. In fact, the electrochemical process takes place in heterogeneous system containing two liquid phases—IL saturated with $TiCl_4$ and vice versa. There are no theoretical studies of this case in the literature yet.

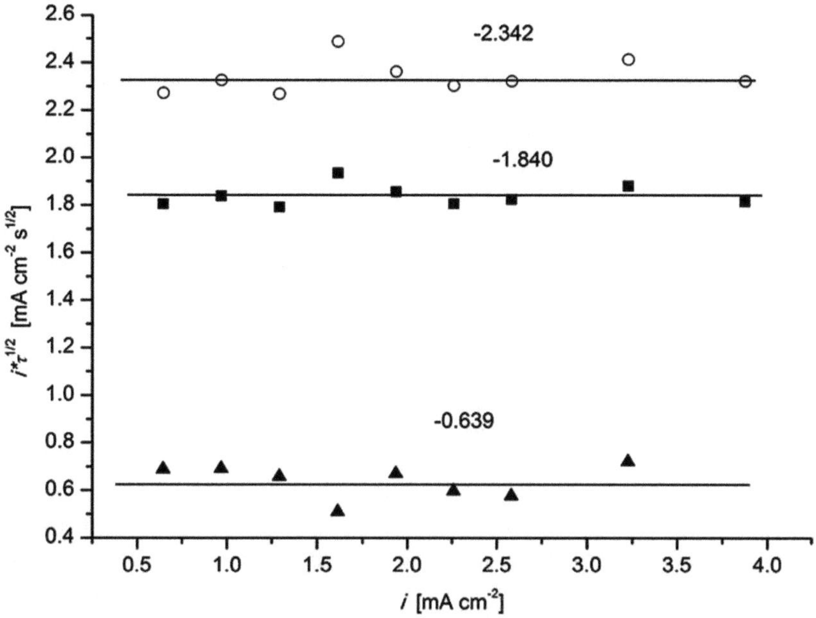

Fig. 6.19 Transition time analysis in terms of Sand equation for three consecutive cathodic waves: (*filled triangle*) peak I at 0.03 V; (*filled square*) peak II at −0.4 V and (*open circle*) peak III at −0.9 V

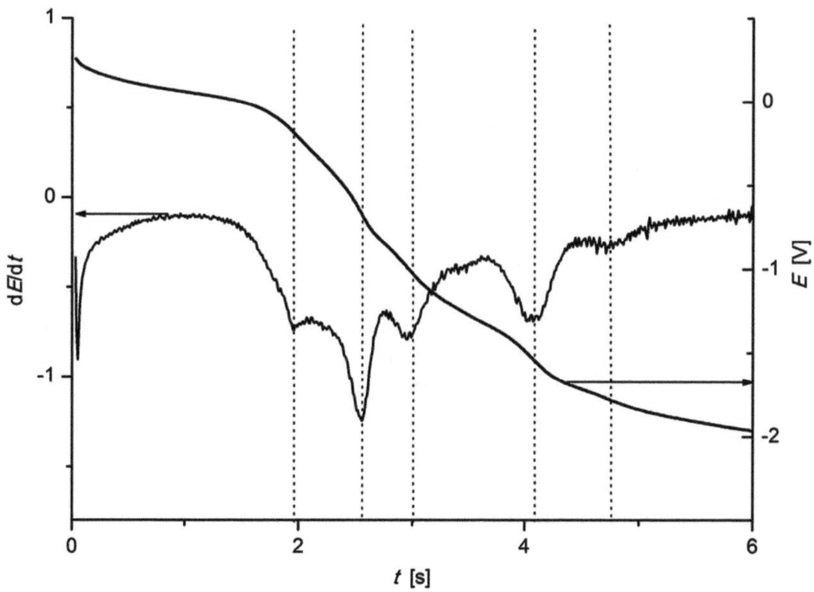

Fig. 6.20 Potential response to a current step (−3.0 mA cm^{-2}) and its first derivative for $C_{TiCl_4} = 0.2$ mol L^{-1}

As follows from the results, the overall electrochemical reduction in the system $TiCl_4$–$BMMImBF_4$ can be represented in terms of general scheme of stepwise reaction mechanism (1.9), which, though simplified, remains very good approximation for the processes in IL media.

In diluted solutions, the Ti(II) intermediate is more stable, and also the disproportionation kinetics of Ti(I) can be detected. Total four-electron reduction to Ti(0) is observed with the formation of Ti(III) and Ti(I) as the relatively stable intermediates.

6.4 Solutions of TiF_4 in 1-Butyl-2,3-Dimethylimidazolium Tetrafluoroborate

An attempt has been made [17] to summarize the results on the electrochemical behaviour of $TiCl_4$ in disubstituted imidazolium-based ionic liquids with bis (trifluoromethylsulfonyl) imide (Tf_2N) anion. The authors [17] came to very interesting conclusion: the reduction of Ti(IV) to Ti metal is essentially impossible in the presence of chloride ions because of the low solubility of the titanium chloride intermediates, which deposit on the cathode in the form of non-stoichiometric halides instead of elemental Ti. Thus, in fact, the electrochemical reduction process of titanium (IV) in these ionic liquids was implicitly recognised to proceed in three-phase system: electrolyte–film–metal.

Now we know that the film can be reduced to pure metal at the interface film–electrode, following a solid state electrochemical reduction mechanism, as it was described in Chaps. 1 and 4. Let us remember that the preconditions for the deposition of the pure metal in such system are the following ones: (1) predominant conductivity of the film must be of ionic nature and (2) the anions rather than the cations must ensure the charge transfer across the film. In that case, the solid film cations can be reduced to the phase with metal type of conductivity. In this case, negative excess charge at the inner boundary will be transported across the film into the electrolyte solution by anions.

Evidently, the bulky Cl^- anions cannot be efficient charge carriers in the solid film. The replacement with more mobile F^- anions can bring the possibility of reducing the film to Ti metal. In order to check this possibility, the electrochemical reduction of Ti (IV) in TiF_4–$BMMImBF_4$ ionic liquids has been studied in [10]. As before, the main electrochemical techniques were chosen to be CV and CP methods.

The mixtures of $BMMImBF_4$ with TiF_4 (range 2–35 mol%) were melted in an inert atmosphere of argon (99.999 %) in an oil bath with stirring until the transparent liquids were formed. The homogenisation takes 15–20 min and the total dissolution of TiF_4 was achieved at 80–85 °C. After cooling, the ionic liquids were liquid at room temperatures. As the concentration of TiF_4 reached 20 mol% (approx. 1 mol L^{-1}), the gas evolution was observed at the temperatures around 80 °C most probably due to the chemical reaction in the IL. Some information about the nature of this reaction was obtained by FTIR spectra measurements of the compositions.

As in the case of $TiCl_4$, the observed CV patterns depend essentially on the concentration of TiF_4 in the electrolyte (Fig. 6.21).

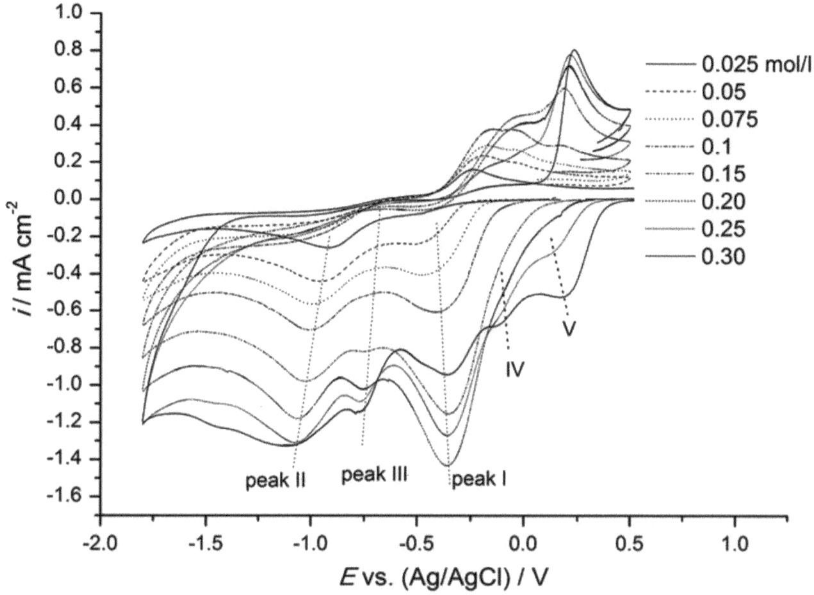

Fig. 6.21 Cyclic voltammograms of $BMMImBF_4$–TiF_4 liquid mixtures at various concentrations of TiF_4. Sweep rate 100 mV s^{-1}, temperature 65 °C

Up to ~0.1 mol L^{-1}, one can observe the typical sequence of two reduction waves (at −0.3 to −0.4 and −0.9 to −1.1 V vs. Ag/AgCl), most probably, corresponding to Ti(IV) $\xrightarrow{e^-}$ Ti(III) $\xrightarrow{e^-}$ Ti(II) stepwise reduction process. The current values for the first peak increase linearly as the concentration of TiF_4 increases (area 1 in Fig. 6.22).

The CV response becomes more complicated with the further increase of the TiF_4 concentration. An additional reduction peak (III in Fig. 6.21) appears at about −0.75 V; the values of the first peak currents are increasing up to 0.2 mol L^{-1} TiF_4, but at higher concentrations the peak current of the peak I decreases (Fig. 6.22).

Further on, as the concentration of TiF_4 becomes higher than 0.2 mol L^{-1}, two more reduction waves appear at less negative potentials in the cathodic part of the cyclic voltammograms (IV and V, Fig. 6.21).

In order to explain this complex reduction pattern observed at higher concentrations of TiF_4, chemical reactions between TiF_4 and $BMMImBF_4$ were assumed, resulting in the appearance of additional species that can be reduced together with the titanium (IV) compounds. In order to verify this assumption, FTIR spectra of TiF_4–$BMMImBF_4$ at different concentrations of TiF_4 were analysed [10].

The FTIR spectra of solid TiF_4, pure $BMMImBF_4$ and TiF_4–$BMMImBF_4$ mixtures with different content of TiF_4 are presented in Figs. 6.23 and 6.24.

In the FTIR spectrum of $BMMImBF_4$, the absorption bands at 1,028 and 1,045 cm^{-1} are assigned to the ν_3 stretching vibrations and the band at 666 cm^{-1} to the ν_4 stretching vibration of BF_4^- anions [18]. As follows from the data in Fig. 6.3, their intensities gradually decrease as the content of TiF_4 in the melt increases. Instead, the new bands at 555, 579, 682, 717, 840 and 887 cm^{-1} appear in the spectra and their intensities rise

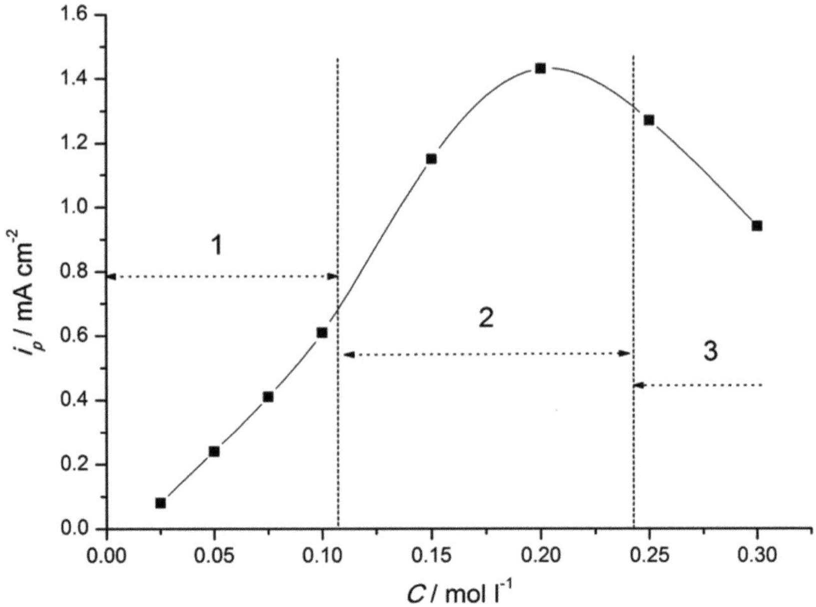

Fig. 6.22 Current of peak I (at –360 mV, Fig. 6.21) vs. concentration of TiF_4

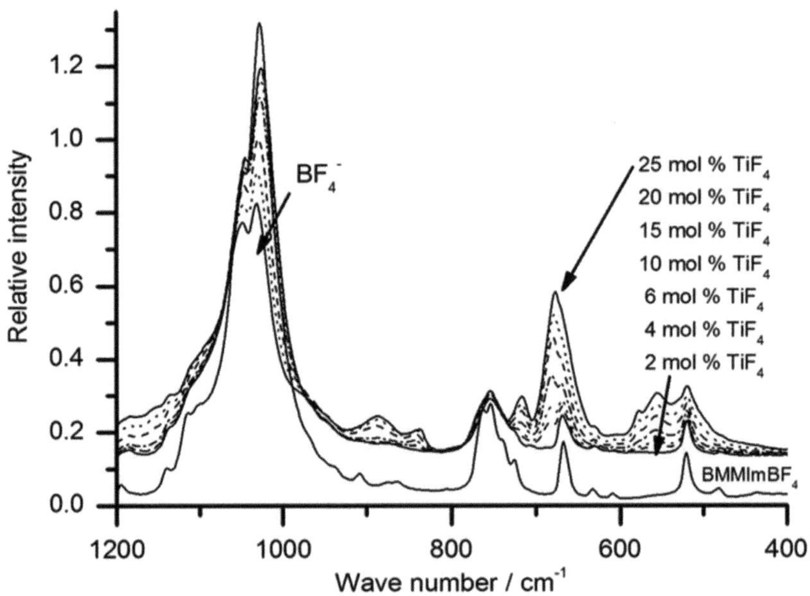

Fig. 6.23 FTIR spectra of $BMMImBF_4$ and the mixtures $BMMImBF_4$–TiF_4 at various concentrations of TiF_4 (2, 4, 6, 10, 15, 20 and 25 mol%), 20 °C

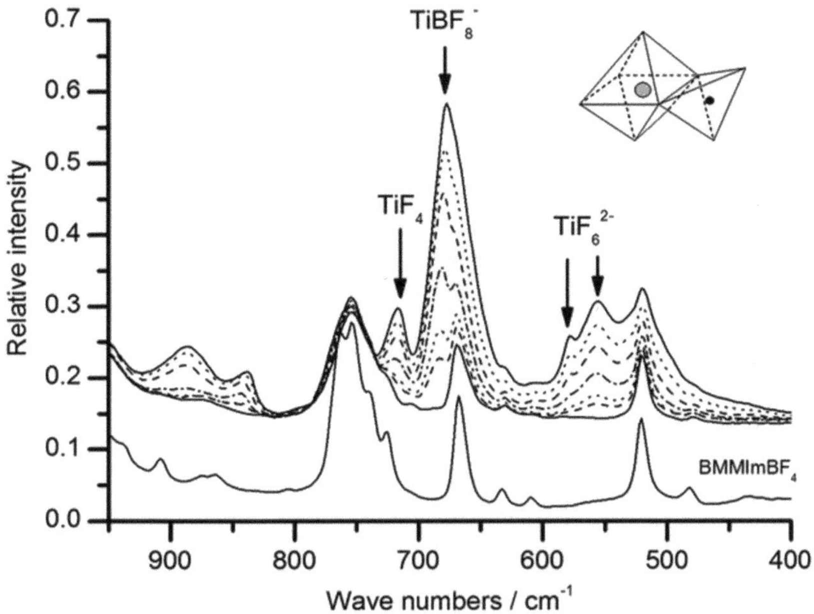

Fig. 6.24 The FTIR spectra of $BMMImBF_4$ and the mixtures $BMMImBF_4$–TiF_4 (Fig. 6.23) in enlarged scale

as the concentration of TiF_4 increases. The comparison of the data obtained from the FTIR spectra of TiF_4 and K_2TiF_6, and also from the spectra of TiF_4 and K_2TiF_6 in the eutectic mixture NaF–KF, allows to identify the presence of TiF_4 (717 cm^{-1}, ν_3 (F_2)) [19] and TiF_6^{2-} complexes (555 cm^{-1} and 579 cm^{-1} sh, ν_3 (F_{1u})) in the system [20]. The most pronounced band at 677 cm^{-1} is attributed to the formation of the hetero-nuclear complexes between BF_4^- and TiF_4, sharing a common edge, with formation of the $TiBF_8^-$ units.

The most plausible explanation for the observed FTIR spectra can be given in terms of two chemical reactions:

$$2BF_4^- + TiF_4 \rightarrow TiF_6^{2-} + 2BF_3 \uparrow \tag{6.7}$$

$$BF_4^- + TiF_4 \rightarrow TiBF_8^- \tag{6.8}$$

The reaction (6.7) accompanied by BF_3 gas evolution is intensified as the concentration of TiF_4 in the IL increases and with the temperature increase. That is confirmed by the increase of the intensity of the band at 555 cm^{-1} assigned to the TiF_6^{2-} anions. Simultaneously, the intensity of the bands at 1,028 $cm^{-1}$and 1,045 cm^{-1} of BF_4^-anions decreases. That can be caused first of all by the gas evolution according to the reaction (6.1) and the participation of BF_4^- in the formation of the complexes according to the reaction (6.8). The gas evolution becomes distinctly visible in concentrated solutions (25–35 mol%) at temperatures above 130 °C. Up to 15–20 mol% TiF_4, the solutions

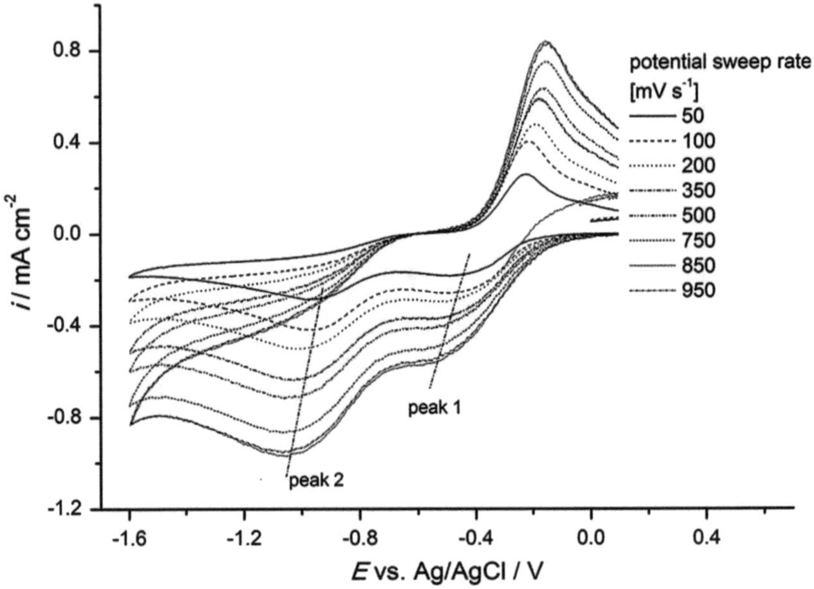

Fig. 6.25 Cyclic voltammograms of 0.05 mol L^{-1} TiF_4 solution at different potential sweep rates at 65 °C

are sufficiently stable if the temperature does not exceed ~130 °C. The reaction (6.8) does not manifest itself for low concentrated solutions (area 1 in Fig. 6.22) in the temperature range up to 180 °C. Thus, the electrolytes with the content of TiF_4 up to ~0.1 mol L^{-1} can be considered as stable enough for electrochemical interpretation. Further electrochemical studies were focused on the electrolytes with concentration of $TiF_4 < 2$ mol% ($\approx$0.1 mol L^{-1}).

Only two reduction waves are observed both in CV (Fig. 6.25) and in CP (Fig. 6.26) curves at 0.05 mol L^{-1} ($\approx$1 mol%) TiF_4 in the temperature range of 65–180 °C.

As follows from the CV data (Fig. 6.25), the peak potential depends on the scan rate, which indicates the irreversibility of the first reduction step. The irreversibility also manifests itself in the width of the observed peak.

The first reduction wave is similar to that observed for the solutions of $TiCl_4$ in the azide melt (see Sect. 6.2). Analysis of the peak currents in terms of the $i_p - v^{1/2}$ plot (Fig. 6.27) is consistent with the theory of Matsuda and Ayabe [14]. We can observe three regions: reversible process—scan rates lower than 0.1 V s^{-1} (area 1), irreversible—scan rates higher than 0.5 V s^{-1} (area 2) and quasi-reversible—process in between (area 3).

Thus, the first one-electron reduction step is not reversible. Another electron is consumed at the second step, which follows from the ratio of the chronopotentiometric transition times approximately equal to $i\tau_2^{1/2}/i\tau_1^{1/2} \cong 2$ (Fig. 6.28).

The product $i\tau_2^{1/2}$ corresponding to the second reduction wave becomes dependent on the current value at the lowest temperature 65 °C (Fig. 6.29).

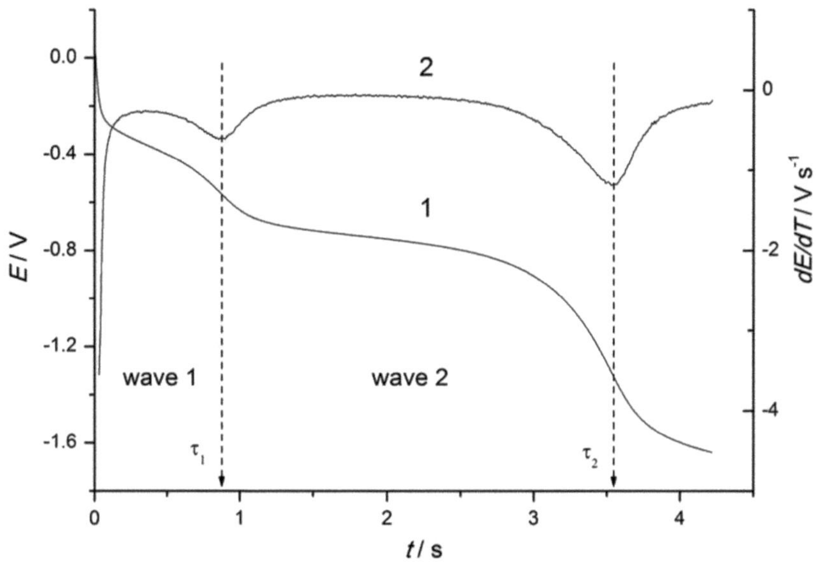

Fig. 6.26 Typical chronopotentiometric curve (1) and its derivative (2), used for determination of transition times: $t = 120\ °C$, $C_{TiF4} = 0.05\ mol\ L^{-1}$, $i = 1.62\ mA\ cm^{-2}$

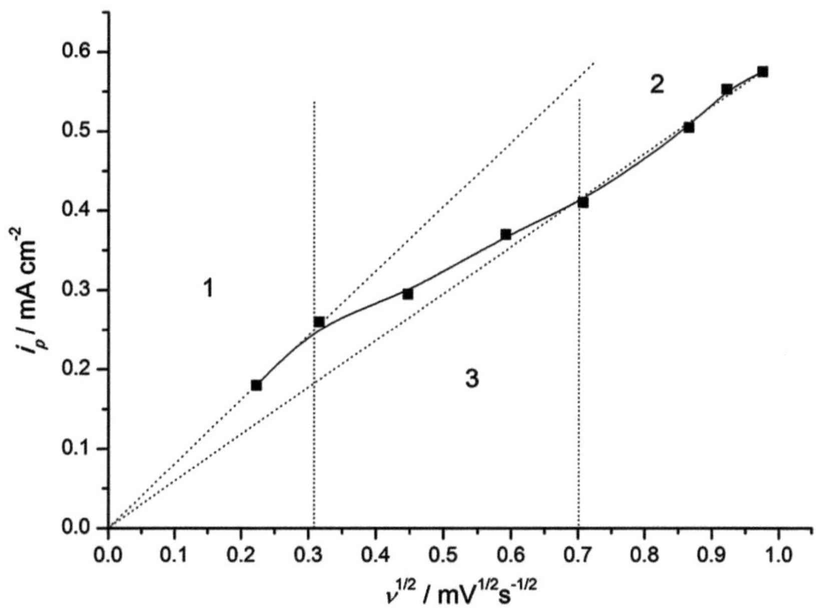

Fig. 6.27 Dependence of the peak currents (for the *peak 1*, Fig. 6.25) on the square root of the potential sweep rate

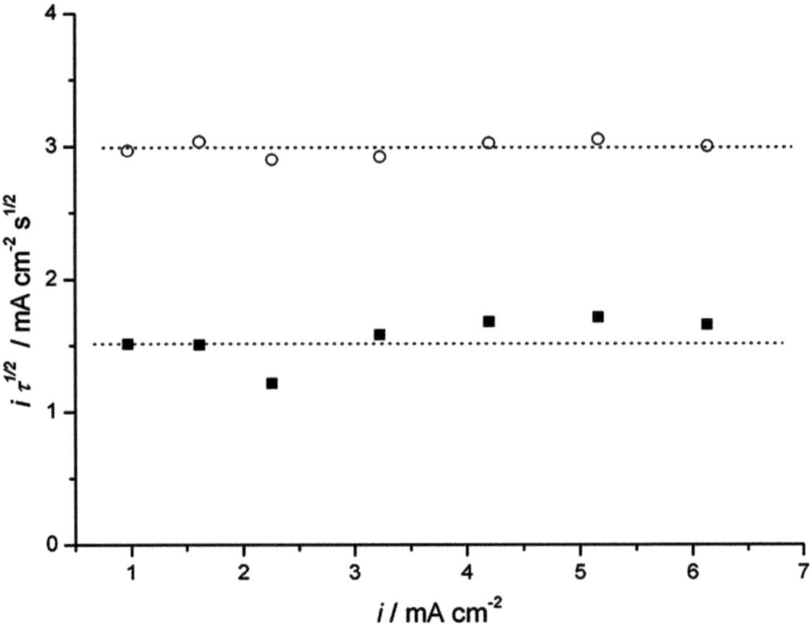

Fig. 6.28 Transition time analysis for CP data: $t = 120\ °C$, $C_{TiF_4} = 0.05$ mol L^{-1} (*filled square*, wave 1; *open circle*, wave 2 in Fig 6.26)

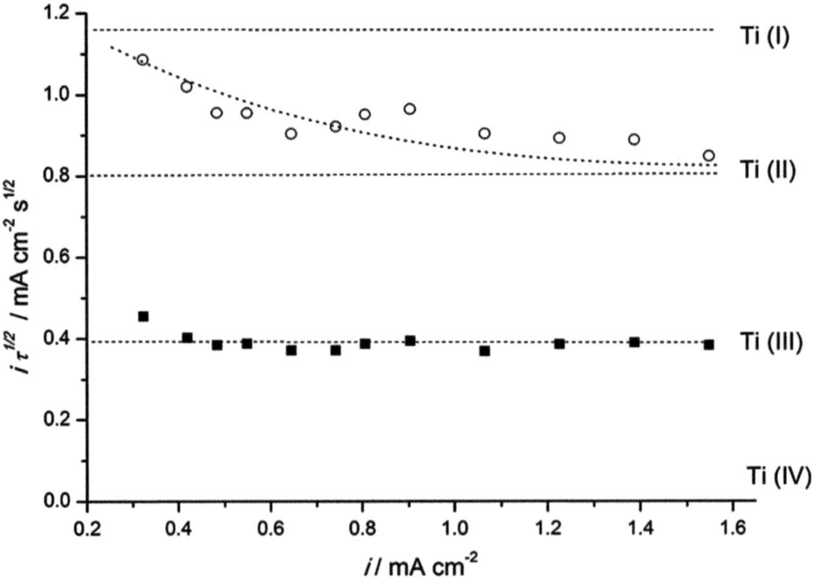

Fig. 6.29 Transition time analysis for two reduction waves of Ti(IV) in 0.05 mol L^{-1} TiF_4 solution at 65 °C

The plot of $i\tau_2^{1/2}$ vs. i has a negative slope. According to the general rule (see Chap. 3), it gives an evidence of the replenishment of the depolarizer by subsequent disproportionation of the product. Then, the reaction mechanism of the second step should be considered as follows:

$$\mathrm{Ti(III)} \xrightarrow{e^-} \mathrm{Ti(II)}; \quad 2\mathrm{Ti(II)} \rightarrow \mathrm{Ti(III)} + \mathrm{Ti(I)} \tag{6.9}$$

According to Eq. (6.9), the overall process at low current densities should correspond to the formation of Ti(I) as a final product. Meanwhile, the chemical reaction according to the scheme (6.9) can be neglected at high current densities, and thus, Ti(II) should be the reduction product. That is, one can expect $\lim\limits_{i\to 0}(i\tau_2^{1/2}/i\tau_1^{1/2}) \approx 3$ and $\lim\limits_{i\to\infty}(i\tau_2^{1/2}/i\tau_1^{1/2}) \approx 2$. The data in Fig. 6.29 demonstrate consistency of the suggestion.

Since the mechanism (6.9) was already theoretically analysed by Fischer and Dracka [16], the data of Fig. 6.29 were represented in terms of their equation:

$$i\tau^{1/2} = C_0 F\pi^{1/2}D_A^{1/2} - \left(\frac{3}{16}\right)^{1/3}\pi^{1/2}D_A^{-1/2}D_B^{2/3}F^{1/3}k^{-1/3}i^{2/3}, \tag{6.10}$$

where A and B are Ti(III) and Ti(II), respectively, and k is the rate constant of the chemical reaction. The plot of the data is shown in Fig. 6.30.

After linear fit of the data in Fig. 6.9, the rate constant was calculated from the slope as $k = (8.7 \pm 3.8) \times 10^5\ \mathrm{cm^3\ mol^{-1}\ s^{-1}}$. It is about a factor of two lower than the disproportionation rate constant of Ti(II) for $TiCl_4$ solutions in $BMMImBF_4$ (see above).

The comparison of the mechanism of TiF_4 reduction to that one of $TiCl_4$ in $BMMImBF_4$ (Sect. 6.3) demonstrates that the non-stationary reduction behaviour of Ti(II) is qualitatively similar.

The values of the diffusion coefficients increase rapidly as the temperature increases. Figure 6.31 presents the Arrhenius plot of the data in Table 6.1.

Activation energy of the diffusion transport calculated from the data of Fig. 6.10 is equal to 38.1 ± 0.7 kJ, which is common for diffusion processes. This value, however, seems to be not precise because of essential non-linearity of the plot—the point at the lowest temperature deviates significantly from the expected position on the linear fit line. That is why the activation energy of diffusion transport was recalculated using the data on CV peak currents at different temperatures.

Arrhenius plot of the first peak current logarithm is shown in Fig. 6.32. Again, the value at the lowest temperature deviates, but other points fit the straight line perfectly well.

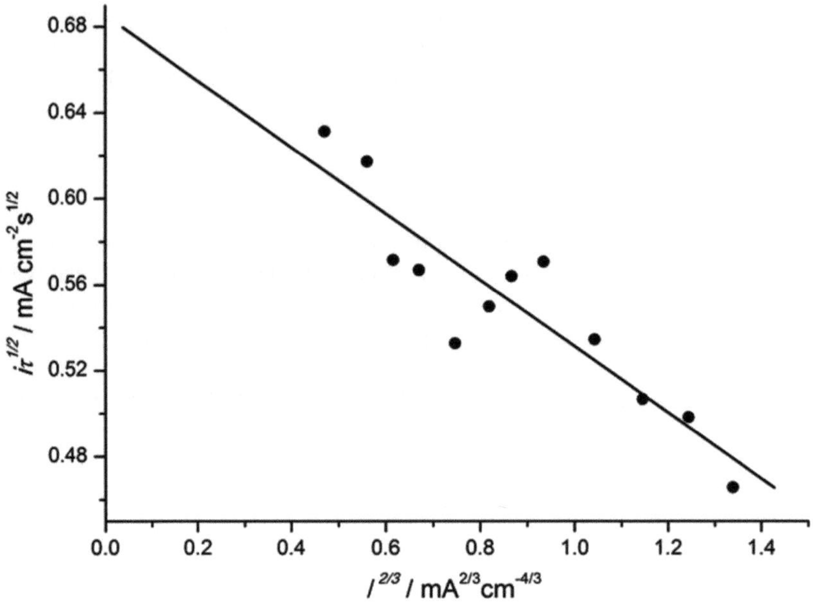

Fig. 6.30 Plot of the data for the second wave of Fig. 6.29 in terms of Eq. (6.10). The transition time here is calculated as follows: $\sqrt{\tau} = \sqrt{\tau_2} - \sqrt{\tau_1}$

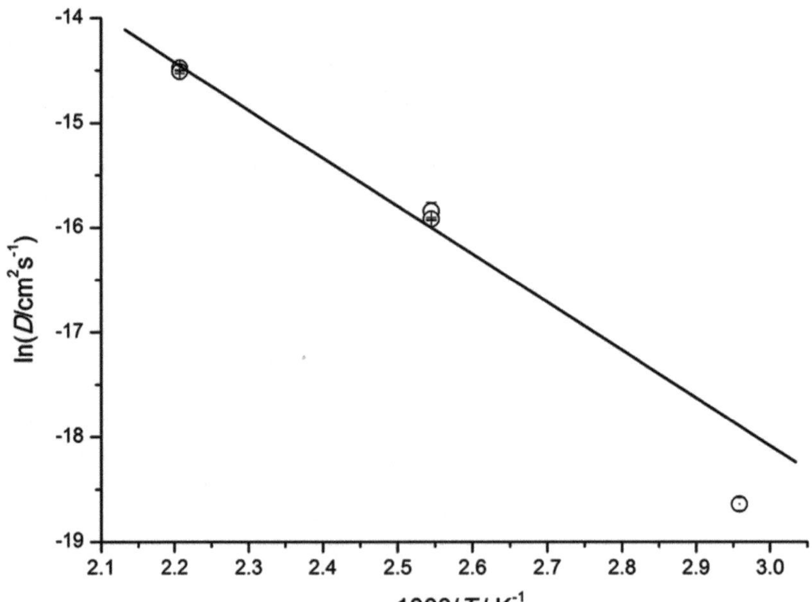

Fig. 6.31 Temperature dependence of $D_{\mathrm{Ti(IV)}}$ in TiF_4–$BMMImBF_4$ solutions according to chronopotentiometric data $C_{\mathrm{TiF_4}} = 0.05$ mol L^{-1}

Table 6.1 Diffusion coefficients of Ti(IV) species in the system calculated from chronopotentiometric data by Sand equation

T (°C)	$D_{Ti(IV)}$ ($cm^2\ s^{-1}$)
65	$(0.80 \pm 0.03) \times 10^{-8}$
120	$(13.2 \pm 0.5) \times 10^{-8}$
180	$(51.2 \pm 0.7) \times 10^{-8}$

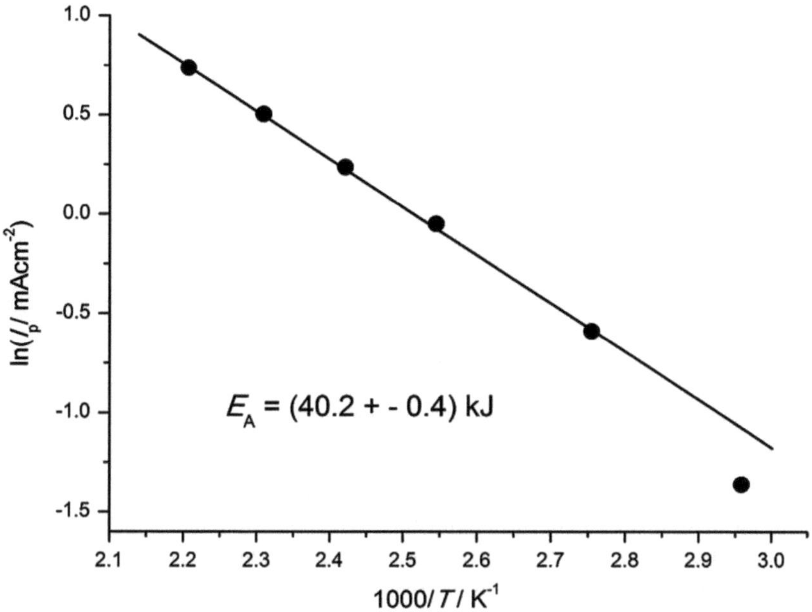

Fig. 6.32 Arrhenius plot of the peak currents for peak I (−450 mV)

Provided Ševčik–Randles equation is valid for the first reduction peak, the logarithmic expression, if sweep rate and concentration are constant, can be written as

$$\ln i_p = \frac{1}{2} \ln D + \text{const} \tag{6.11}$$

Hence, the slope of $\ln i_p$ vs. the reverse temperature should be equal to $-E_A/2R$, where E_A is the activation energy of diffusion. The value of E_A, calculated from the data in Fig. 6.32, is equal to 40.2 ± 0.4 kJ and is consistent with the activation energies of viscous flow and conductivity of pure $BMMImBF_4$ [11], indicating similarity of the transport mechanisms.

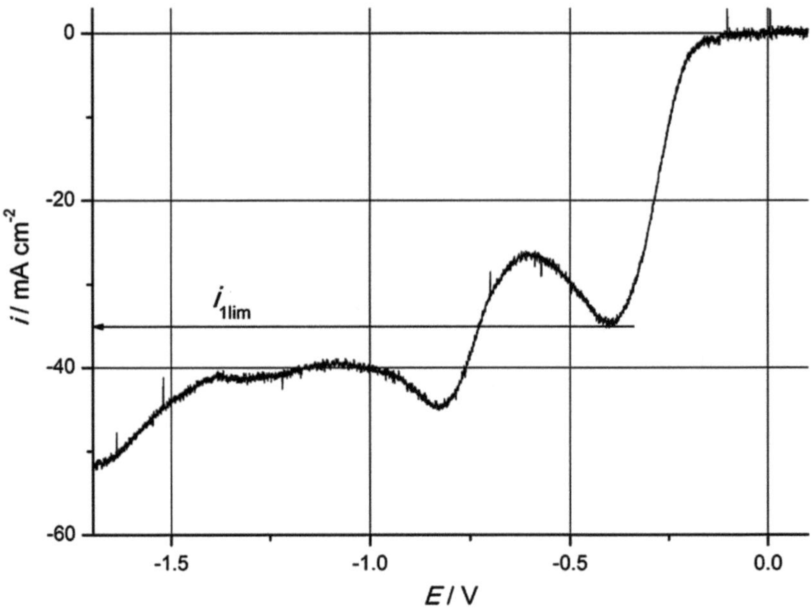

Fig. 6.33 Polarization curve at the potential sweep rate 0.5 mV s^{-1}, $C_{TiF_4} = 0.05$ mol L^{-1}, $t = 65$ °C

Thus, in the low concentration region (1 in Fig. 6.22) stepwise reduction of Ti(IV) is observed with formation of the Ti(III) and Ti(II) intermediates. The latter is able to form Ti(I) intermediate according to the disproportional reaction [scheme (6.9)]. This is consistent with general scheme of a many-electron reduction process (1.9). In this case, the general mechanism (1.9) can be adjusted as follows:

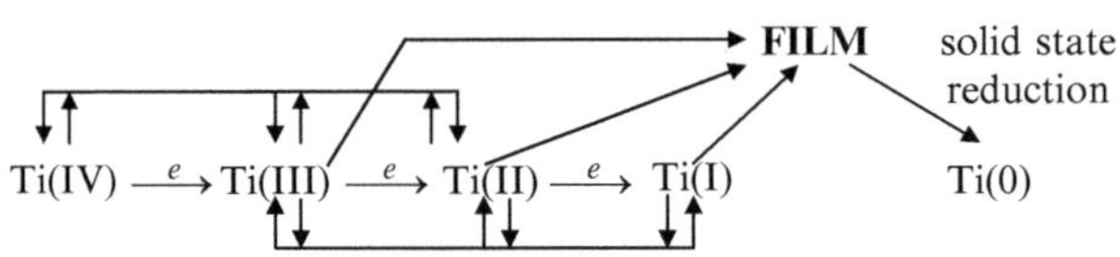

(6.12)

The contours in scheme (6.12) denote disproportionation reactions which could proceed both at equilibrium and in kinetic conditions. The straight arrowed lines show the possibility of titanium intermediate valency compounds to deposit at the electrode surface and to form an electrochemicaly active film system. The experimental proof of such a possibility follows from the analysis of stationary polarization curve shown in Fig. 6.33.

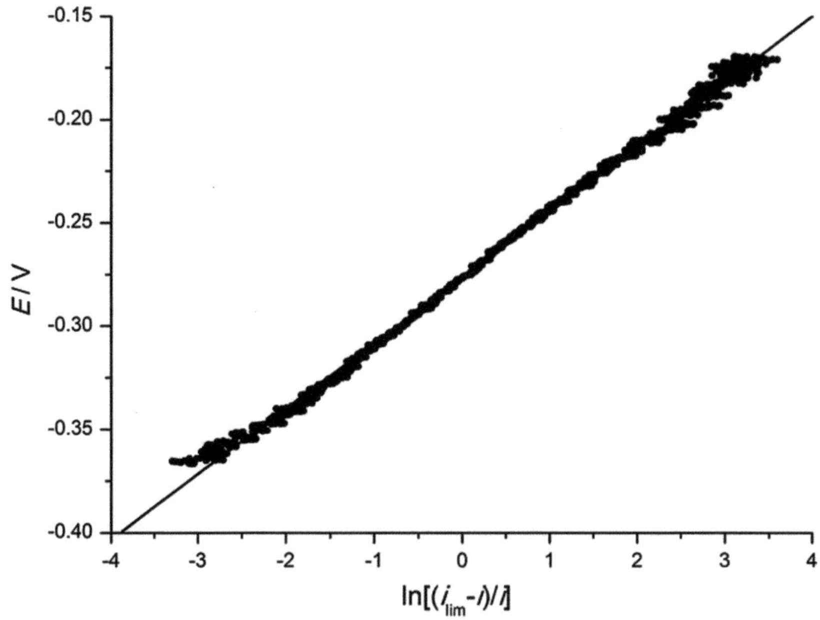

Fig. 6.34 Semi-logarithmic plot of the first reduction wave Fig. 6.13 in terms of Eq. (6.13)

As follows from the analysis given above (Fig. 6.27), the reduction process at the potential sweep rate lower than 0.1 V s^{-1} is reversible. This means that the stationary polarization curve should be described by the Heyrovský–Ilkovich equation type for a polarographic wave with a limiting current. However, Fig. 6.33 shows no common limiting current plateau on LSV curves even at very low potential sweep rates. Instead, the current descends after reaching the maximum value. Meanwhile the shape of the curve looks at the first sight like a classical reversible polarographic wave. The situation is quite similar to that described in Chap. 3 for the reduction of Ge(IV) in oxyfluoride melts (see Fig. 3.7).

Indeed, the first wave, plotted in semi-logarithmic coordinates, is a straight line (Fig. 6.34) in accordance with the Heyrovský–Ilkovich equation for a reversible polarographic curve:

$$E = E_{1/2} + \frac{RT}{nF} \ln \frac{i_{\lim} - i}{i} \tag{6.13}$$

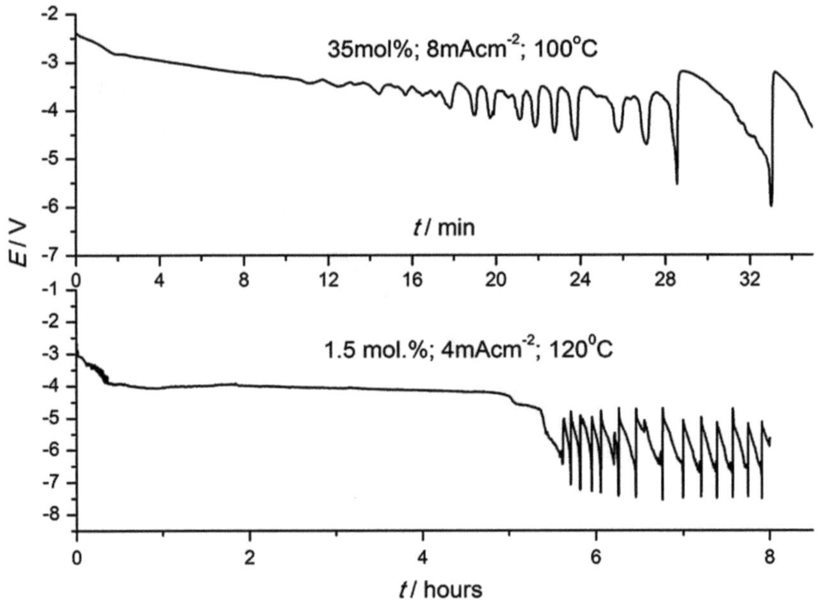

Fig. 6.35 Change of the cell voltage in time during the galvanostatic electrolysis of TiF_4 solutions in $BMMImBF_4$

The pre-logarithm coefficient calculated from the plot, Fig. 6.34, is equal to 32 mV, which is in quite a good agreement with the theoretical value of 29 mV for one-electron process at 65 °C. Thus, the decrease of the current after going through a maximum cannot be explained by non-stationary diffusion restrictions as in the case of common LSV of Ševčik–Randles type. The reason for the current decrease is the increase of the ohmic resistance due to a deposition of the reduction product on the electrode surface.

The formation of a certain deposit at the electrode can be visually seen as the electrode surface becomes darker after the moment when current attains its maximum.

The second stationary wave exhibits similar maximum (Fig. 6.33). After that, at more negative potentials than −1.2 V, the reduction is related (at least partly) to the solid state process. Hence, there is a possibility of reducing the Ti(IV) to Ti metal in this system by means of long-time electrolysis. This possibility was verified in DC electrodeposition experiments, which were carried out for 3–8 h in galvanostatic conditions at 180 °C.

During the electrolysis, the cell voltage increases gradually in time, thus indicating the accumulation of sparingly soluble intermediates on the electrode (Fig. 6.15).

After definite time (from 10 min up to several hours depending on the current, composition and temperature), the process becomes unsteady which causes the non-linear oscillations of voltage (Fig. 6.35). As we already know (see Chaps. 4 and 5), such behaviour is often observed in electrochemical film systems with mixed ion–electron conductivity of the film.

Fig. 6.36 Optical microscope images of electrolytic deposits of titanium on Pt wire (*left*) and on the Pt chip

It was found that the product of electrolysis at concentrations of $TiF_4 > 0.1$ mol L^{-1} consists mainly of a fine black powder which is very difficult to handle because of the poor adhesion to the surface of the electrode and rapid oxidation in the presence of the traces of oxygen. Thin (up to 2 μm) smooth coherent deposits were obtained in the electrolytes at lower (up to 0.1 mol L^{-1}) concentrations of TiF_4 at 180 °C (Fig. 6.36).

The EDX elemental analysis of the deposit surface (Fig. 6.37) showed Ti as the main component.

The dark deposit (Figs. 6.36 and 6.37) becomes white in contact with ambient air due to the oxidation of the electrolytically deposited, very fine grained Ti metal which has a very high chemical activity. As it follows from the Raman spectrum (Fig. 6.38), the surface of even freshly prepared and handled under argon Ti deposits is immediately covered by the thin film of titanium dioxide during Raman spectrum acquisition under ambient air.

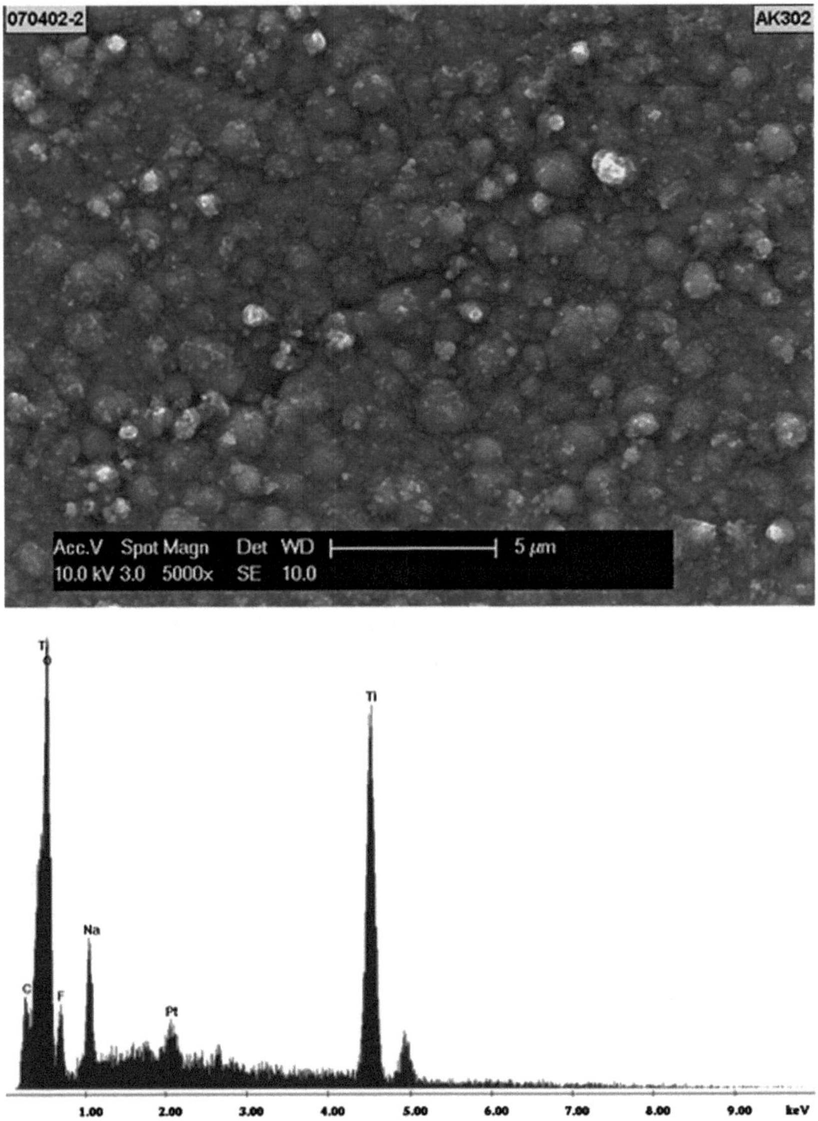

Fig. 6.37 ESEM image of the of titanium deposit at ×5,000 magnification in secondary electrons mode (*above*) and the EDX analysis of the surface layer (*below*)

6.5 Conclusions

Except for the azide IL where only shallow reduction of Ti(IV) is possible, the non-stationary reduction mechanism of Ti(IV) in both a titanium chloride and fluoride–1-butyl-2,3-dimethyl imidazolium tetrafluoroborate electrolyte is a sequence of one-electron steps with the formation of the Ti(III), Ti(II) and Ti(I) low valence

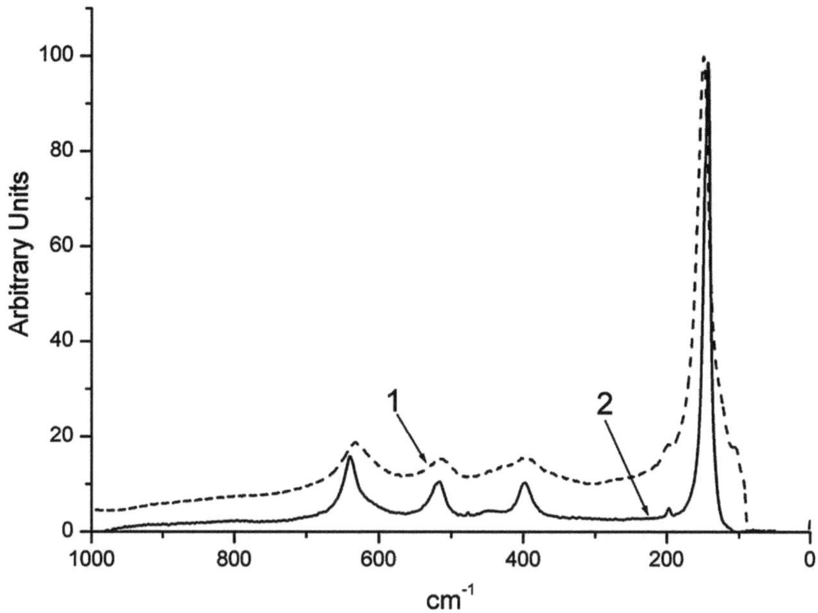

Fig. 6.38 Raman spectra of the deposits (1) and TiO_2 anatase from the database (2)

compounds, which are poorly soluble in the electrolyte. This fact is in agreement with the general mechanism developed for the reduction of polyvalent method compounds in high-temperature molten salt electrolytes.

At the conditions of long-term stationary electrolysis, the intermediate valency titanium compounds are deposited at the electrode surface forming a three-phase electrode/film/metal system, which is also very common in high-temperature melts. Due to the presence of mobile F^- anion, the reduction of the film's substances to Ti metal becomes possible in long-term electrolysis process in the system TiF_4–$BMMImBF_4$.

References

1. Wasserscheid P, Welton T (2004) Ionic liquids in synthesis. Wiley-VCH, Weinheim
2. Walden P (1914) Bull Acad Imper Sci (St Petersburg, 1800) 8:405
3. Bonhote P, Dias AP, Papageorgiou N, Kalyanasundaram K, Grätzel M (1996) Inorg Chem 35: 1168–1178
4. McEwen AB, Ngo HL, LeCompte K, Goldman JL (1999) J Electrochem Soc 146:1687–1695
5. Nishida T, Tashiro Y, Yamamoto M (2003) J Fluorine Chem 120:135–142
6. Ohno H (2005) Electrochemical aspects of ionic liquids. Wiley, New York, NY
7. Gifford PR, Palmisano JB (1987) J Electrochem Soc 134:610–614
8. Andriyko Y, Nauer GE (2007) Electrochim Acta 53:957
9. Andriyko Y, Fastner U, Kronberger H, Nauer GE (2007) Z Naturforsch 62a:529–538
10. Andriyko Y, Andriiko A, Babushkina OB, Nauer GE (2010) Electrochim Acta 55:1081–1089
11. Andriyko YO, Reischl W, Nauer GE (2009) J Chem Eng Data 54(3):855

12. Xiao L, Johnson KE (2003) J Electrochem Soc 150(6):E307
13. Bard AJ, Faulkner LR (2001) Electrochemical methods: fundamentals and application, 2nd edn. Wiley, New York, NY
14. Matsuda H, Ayabe Y (1955) Z Elektrochem 59:494
15. Trulove P, Mantz RA (2003) In: Wasserscheid P, Welton T (eds) Ionic liquids in synthesis. Wiley-VCH, Weinheim, p 103
16. Fischer O, Dračka O (1959) Collect Czhech Chem Commun 24:3046
17. Endres F, Zein El Abedin S, Saad AY, Moustafa EM, Borissenko N, Price WE, Wallace GG, MacFarlane DR, Newman PJ, Bund A (2008) Phys Chem Chem Phys 10:2189
18. Goubeau J, Bues W (1952) Z Anorg Allg Chem 268:221
19. DeVore TC, Gallaher TN (1985) J Chem Phys 82:2512
20. Forest IW, Lane AP (1976) Inorg Chem 15:265

Afterword

Hereby we finish our narrative on the electrochemistry of many-electron electrochemical processes.

We tried to demonstrate the application of the stepwise mechanism of discharge taking into consideration the intervalent reactions, which accounts for the formation of electrode film systems, which are the active participants of the electrolysis process.

We started from the solid thermodynamic background of equilibrium, then went through and finally came to unsteady field of non-linear dynamics of electrochemical systems. The results in this field are scarce; apart from the above, only a few works can be mentioned related to aqueous electrochemistry. This wild land is still waiting for explorers.

A.A. Andriiko et al., *Many-electron Electrochemical Processes*,
Monographs in Electrochemistry, DOI 10.1007/978-3-642-35770-1,

About the Authors

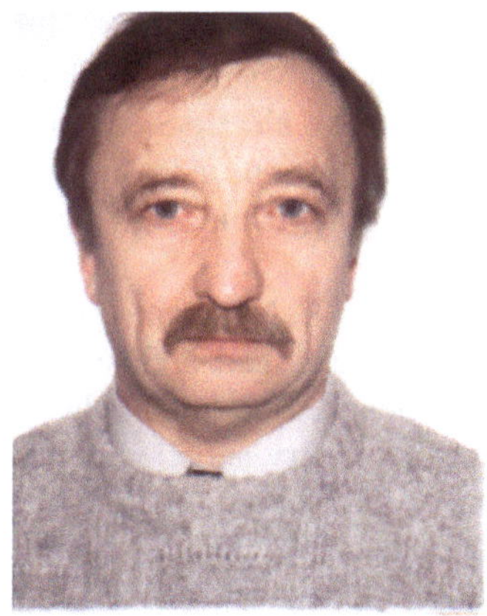

Aleksandr A. Andriiko (born 1951) has been a professor and head of Chair of General and Inorganic Chemistry of the National Technical University of Ukraine "Kyiv Polytechnic Institute" (NTUU "KPI") since 2003. He obtained his diploma in Chemical Engineering in 1974 and worked in the Institute of General and Inorganic Chemistry (IGICH) of the National Academy of Science of Ukraine since then and till 2003. He received his degrees of PhD (Kandidate) in 1981 and DSc in 1993 from IGICH for the works in field of electrochemistry of molten fluorides.

He worked as scientific advisor in Republic of Korea (1998) and as visiting Professor at the Centre of Electrochemical Surface Technology (CEST) in Wiener-Neuschtadt, Austria (Summer 2008).

He has carried out researches in various fields of inorganic chemistry (coordination chemistry of 3d-metals, high-temperature chemistry of molten fluorides), physical chemistry (phase equilibria in molten salt systems), electrochemistry (thermodynamics of metal–electrolyte interface, electrodeposition of refractory and other metals from molten salts, lithium batteries and active materials for them) and, more recently, nanochemistry of inorganic oxide materials.

The results of his research activities were published in more than 120 papers in refereed journals. Also he is co-author of 24 inventions.

A.A. Andriiko et al., *Many-electron Electrochemical Processes*,
Monographs in Electrochemistry, DOI 10.1007/978-3-642-35770-1,

Yuriy Andriyko works now as a senior researcher at the Centre of Electrochemical Surface Technology in Austria.

In 2002 he got his bachelor's degree in Chemical Technology and Engineering at the Technical University of Ukraine "Kyiv Politechnical Institute". At the same University in 2004, he acquired a Master of Science degree in Electrochemistry by defending the Diploma on the topic of "Electrochemical properties of the thermal decomposition products of lithium permanganate". Further on, he continued his studies at the Technical University of Vienna, and in the year 2007, he defended his PhD Thesis on the subject "Electrochemistry of Ti in ionic liquids".

Among his research interests are investigation of the cathode materials for the lithium–ion batteries, physical–chemical properties of low-temperature ionic liquids and electrochemistry of refractory metals in ionic liquids. Recently he has been taking part in the research and optimization on the industrial scale of processes of Zn, Zn–Fe and Zn–Ni alloy deposition.

Gerhard E. Nauer has studied chemistry, physics and classical archaeology at the University of Vienna, Austria, and he specialised in physical chemistry and electrochemistry. Since 1975 he worked at the Institute of Physical Chemistry at the University of Vienna. His research has been focused on battery technology, in situ electrochemical methods, surface technology and corrosion. In 2000 he founded the "K*plus*-Kompetenzzentrum ECHEM für Angewandte Elektrochemie" (a centre of competence in applied electrochemistry), and he was its head until 2008. He is consultant for various national and international companies, especially in the area of "green surface technology", batteries and corrosion. Nauer has more than 150 publications in the field of applied electrochemistry and is author or co-author of 12 patents. Currently he is president of "IG Energieautarkie" (community of interest in energetic autarchy) and member of the board of PFI ("platform innovation management") in Austria.

About the Editor

Fritz Scholz is Professor at the University of Greifswald, Germany. Following studies of chemistry at Humboldt University, Berlin, he obtained a Dr. rer. nat. and a Dr. sc. nat. (habilitation) from that University. In 1987 and 1989 he worked with Alan Bond in Australia. His main interest is in electrochemistry and electroanalysis. He has published more than 280 scientific papers, and he is editor and co-author of the book "Electroanalytical Methods" (Springer, 2002, 2005, 2010, and Russian Edition: BINOM, 2006), co-author of the book "Electrochemistry of Immobilized Particles and Droplets" (Springer 2005), co-editor of the "Electrochemical Dictionary" (Springer, 2008) and co-editor of volumes 7a and 7b of the "Encyclopedia of Electrochemistry" (Wiley–VCH 2006). In 1997 he founded the *Journal of Solid State Electrochemistry* (Springer) and serves as Editor-in-Chief since that time. He is the editor of the series "Monographs in Electrochemistry" (Springer) in which modern topics of electrochemistry are presented. Scholz introduced the technique 'Voltammetry of Immobilized Microparticles' for studying the electrochemistry of solid compounds and materials; he introduced three-phase electrodes to determine the Gibbs energies of ion transfer between immiscible liquids, and currently he is studying the interaction of free oxygen radicals with metal surfaces, as well as the interaction of liposomes with the surface of mercury electrodes in order to assess membrane properties.

A.A. Andriiko et al., *Many-electron Electrochemical Processes*, Monographs in Electrochemistry, DOI 10.1007/978-3-642-35770-1,

Index

A.A. Andriiko et al., *Many-electron Electrochemical Processes*,
Monographs in Electrochemistry, DOI 10.1007/978-3-642-35770-1,

MIX
Papier aus verantwortungsvollen Quellen
Paper from responsible sources
FSC® C105338

If you have any concerns about our products,
you can contact us on
ProductSafety@springernature.com

In case Publisher is established outside the EU,
the EU authorized representative is:
Springer Nature Customer Service Center GmbH
Europaplatz 3, 69115 Heidelberg, Germany

Printed by Libri Plureos GmbH
in Hamburg, Germany